EX-PRODIGY

MY CHILDHOOD AND YOUTH

Other books by Norbert Wiener

Selected Technical Papers, 1964

God and Golem, Inc., 1964

Cybernetics, Second Edition, 1961

Nonlinear Problems in Random Theory, 1958

Extrapolation, Interpolation, and Smoothing of Stationary Time Series, 1949

I am a Mathematician: The Later Life of a Prodigy, 1956

Ex-Prodigy: My Childhood and Youth, 1953

EX-PRODIGY

MY CHILDHOOD AND YOUTH

by

NORBERT WIENER

THE M.I.T. PRESS
MASSACHUSETTS INSTITUTE OF TECHNOLOGY
CAMBRIDGE, MASSACHUSETTS

Copyright © 1953 by Norbert Wiener
All rights reserved

First edition published 1953 by Simon and Schuster, Inc.

First M.I.T. Press Paperback Edition, August, 1964

Library of Congress Catalog Card No. 64-22203
Printed in the United States of America

ISBN: 978-0-262-23011-7 (hc.:alk.paper)
ISBN: 978-0-262-73008-2 (pb.:alk.paper)

« TO MY WIFE »
Under whose gentle tutelage
I first knew freedom

« FOREWORD »

THE AUTHOR *wishes to acknowledge the help he has received from many sides in the writing of this book. First of all, the greater part of the first draft of the manuscript was dictated to my wife in Europe and Mexico in 1951, in such divers places as Madrid, Saint-Jean de Luz, Paris, Thonon-les-Bains, Cuernavaca, and Mexico City. In Mexico, Miss Concepcion Romero of the Instituto Nacional de Cardiologia helped me in the retyping of a revised version of the manuscript. And finally at the Massachusetts Institute of Technology, my secretary, Mrs. George Baldwin, has borne with me through the long work of revision and selection which has been necessary for the production of the finished manuscript. I have been greatly assisted in the final typing of the manuscript by Miss Margaret FitzGibbon, Miss Sally Starck, and Miss Katharine Tyler of the Massachusetts Institute of Technology. As this book has been done almost entirely by dictation, the secretarial assistance I have received from these various sources*

FOREWORD

is a vital and constructive contribution to the production of the book.

I have submitted my manuscript for inspection to many friends, and I wish to thank them for their detailed criticism and their positive and negative suggestions. Besides my wife, who has collaborated with me through the entire book, I wish to mention Dr. Marcel Monnier of Geneva; Mr. F. V. Morley and Sir Stanley Unwin of London; Dr. Arturo Rosenblueth of the Instituto Nacional de Cardiologia in Mexico City, Dr. Moritz Chafetz and Dr. William Osher, also temporarily at the Instituto; Dr. Dana L. Farnsworth, Dean F. G. Fassett, Jr., Professors Georgio de Santillana, Karl Deutsch, Arthur Mann, and Elting E. Morison of the Massachusetts Institute of Technology; Professors Oscar Handlin and Harry Wolfson of Harvard University; and Dr. Janet Rioch of New York. Of all these, I should like particularly to single out Professor Deutsch, who has gone into my work with an amount of detailed criticism which is altogether beyond what I could expect of a friend who has volunteered to read the book.

In the firm of Simon and Schuster, my publishers, Mr. Henry W. Simon has had the job of seeing this book through the press. I wish to express particularly my gratitude for his sensitive criticism and his understanding comments.

NORBERT WIENER

Cambridge, Massachusetts
June, 1952

« TABLE OF CONTENTS »

FOREWORD

INTRODUCTION 3

I   A Russian Irishman in Kansas City 7
II The Proper Missourians 23
III First Remembered Patterns: *1894–1901* 31
IV Cambridge to Cambridge, via New York and Vienna: *June–September, 1901* 48
V   In the Sweat of My Brow: *Cambridge, September, 1901–September, 1903* 58
VI Diversions of a *Wunderkind* 78
VII A Child among Adolescents: *Ayer High School, 1903–1906* 92

TABLE OF CONTENTS

VIII	College Man in Short Trousers: *September, 1906–June, 1909*	102
IX	Neither Child nor Youth	115
X	The Square Peg: *Harvard, 1909–1910*	125
XI	Disinherited: *Cornell, 1910–1911*	143
XII	Problems and Confusions: *Summer, 1911*	157
XIII	A Philosopher Despite Himself: *Harvard, 1911–1913*	164
XIV	Emancipation: *Cambridge, June, 1913–April, 1914*	180
XV	A Traveling Scholar in Wartime: *1914–1915*	204
XVI	Trial Run: Teaching at Harvard and the University of Maine: *1915–1917*	227
XVII	Monkey Wrench, Paste Pot, and the Slide Rule War: *1917–1919*	247
XVIII	The Return to Mathematics	264
EPILOGUE		288
INDEX		299

EX-PRODIGY

MY CHILDHOOD AND YOUTH

« INTRODUCTION »

As THIS book will show, at one period I was an infant prodigy in the full sense of the word, for I entered college before the age of twelve, obtained my bachelor's degree before fifteen, and my doctorate before nineteen. Yet any man who has reached the age of fifty-seven is certainly no longer an infant prodigy; and if he has accomplished anything in life, whatever temporary conspicuousness he may have had as a prodigy has lost all importance in view of the much greater issues of success or failure in his later life.

But the present book does not attempt to be an evaluation of my whole life for better or for worse. It is rather the study of a period in which I underwent a rather unusual and early course of education, and of the subsequent period in which the unevenness and irregularity thus accentuated in my nature had an opportunity to work themselves out in such a way that I could consider myself launched on an active career both as a scholar and as a citizen of the world.

The infant prodigy or *Wunderkind* is a child who has

INTRODUCTION

achieved an appreciable measure of adult intellectual standing before he is out of the years usually devoted to a secondary school education. The word "prodigy" cannot be interpreted as either a boast of success or a jeremiad of failure.

The prodigies whom we generally call to mind are either people like John Stuart Mill and Blaise Pascal, who had proceeded from a precocious youth to an effective adult career, or their antitheses, who have found the transition between early precocity and later effectiveness one which they have become too specialized to make. Yet there is nothing in the word itself which restricts us to these two opposite cases. It is perfectly conceivable that after an especially early start, a child may find a place in life in which he has a good, modest measure of success without having stormed Olympus.

The reason infant prodigies are generally adjudged in terms of immense failure or immense success is that they are somewhat rare phenomena, known by hearsay to the public; and, therefore, the only ones the public ever hears of are those who "point a moral or adorn a tale." There is a tragedy in the failure of a promising lad which makes his fate interesting reading; and the charm of the success story is known to all of us. Per contra, the account of a moderate success following a sensationally promising childhood is an anticlimax and not worth general attention.

I consider this attitude of extremes toward the infant prodigy false and unjustified. In addition to being unjustified, it is in fact unjust; for this feeling of anticlimax which the story of the moderately successful prodigy excites in the reader leads the prodigy to a self-distrust which may be disastrous. It takes an extremely solid character to step down gracefully from the dais of the prodigy onto the more modest platform of the routine teacher or the laboratory floor of the adequate but secondary research man. Thus the child prodigy who is not

INTRODUCTION

in fact a prodigy of moral strength as well must make a career success on a large scale, for want of which he is likely to consider himself a failure, and actually to become one.

The sentimentality with which the adult regards the experiences of a child is no genuine part of a child's attitude toward himself. To the grownup, it is lovable and right for the child to be confused, to be at a loss in the world of adults around him; but it is far from a pleasant experience for the child. To be immersed in a world which he cannot understand is to suffer from an inferiority which has no charms whatever for him. It may be ingratiating and amusing to his seniors to observe his struggles in a half-understood world, but it is no more pleasant for him to realize that his environment is too much for him to cope with than it would be for an adult under the same circumstances.

Our present age is sundered from that of the Victorian by many changes, not least among which is the fact that Sigmund Freud has lived, and that no one can write a book nowadays without being aware of his ideas. There is a great temptation to write an autobiography in the Freudian jargon, more especially when a large part of it is devoted to the very Freudian theme of father-and-son conflict. Nevertheless, I shall avoid the use of this terminology. I do not consider that the work of Freud is so final that we should freeze our ideas by adopting the technical language of what is certainly nothing but the contemporary phase of a rapidly growing subject. Yet I cannot deny that Freud has turned over the stone of the human mind and has shown a great population of pale and emotionally photophobic creatures scuttling back into their holes. However, I do not accept all Freudian dogmas as unquestionable truth. I do not consider that the present vogue of the emotional strip-tease is a wholly good thing. But, let my reader make no mistake: the resemblance between many of the

INTRODUCTION

ideas of this book and certain of the notions of Freud is not purely coincidental; and if he finds that he can carry out the exercise of translating my statements into Freudian jargon, he should consider that I am quite aware of this possibility and have rejected it deliberately.

«I»

A RUSSIAN IRISHMAN IN KANSAS CITY

MY IMPRESSION of the intellectual environment in the first decade of this century is intense and actual, and I have learned much of it as a child sitting in the kneehole of my father's desk while he discussed with his friends the vicissitudes of those times and the facts of all times. Child as I was, I absorbed a real understanding of many things, and my childish point of view is not totally devoid of significance. Those of us who make scholarship our career are often able to take the dispersed and uncorrelated memories of our childhood, covering much which we had not understood at the time they were received, and to build them into an organized and cogent structure.

Today all of us have grown up into an age which, while it may be an age of losses and of decline, has also been an age of new beginnings. In these beginnings the scientist and even the mathematician have had a large share. I have been both a witness of and a participant in them. Thus I may speak of them, not only with the understanding of the participant, but I hope also with something of the normative judgment of the objective critic.

That part of my work which appears to have excited the greatest public interest and curiosity concerns what I have called Cybernetics, or the theory of communication and control, wherever it may be found, whether in the machine or in the living being. I have had the good fortune to say something about these matters. This has not been merely the *aperçu* of a moment. It has its deep roots both in my personal development and in the history of science. Historically it stems from Leibnitz, from Babbage, from Maxwell, and from Gibbs. Within me, it stems from the little I know of these masters and from the way this knowledge has fermented in my mind. Therefore, perhaps, an account of the origins of my predisposition towards these ideas, and of how I came to take them as significant, may be of interest to the others who are as yet to tread my road.

As far as I know, I am about seven-eighths Jewish in ancestry, with one possible German Lutheran great-grandparent on my mother's side. Because I am myself overwhelmingly of Jewish origin, I shall have more than one occasion to refer to Jews and Judaism. Since neither I myself nor my father nor, so far as I know, his father has been a follower of the Jewish religion, I must explain the sense in which I tend to use the word "Jew" and all related words such as "Judaism" and "Gentile" which are given a definition in terms of the master word.

The Jews seem to me primarily a community and a social entity although most of them have been members of a religion as well. Nevertheless, when this religion has begun to offer a less impenetrable barrier to the surrounding community, and when the surrounding community has begun to offer a less impenetrable barrier toward it, there are many factors in the

A RUSSIAN IRISHMAN IN KANSAS CITY

life of those who had adhered to the religion which have continued to follow more or less the original religious patterns. The Jewish family structure is somewhat closer than the average European family structure, and much closer than that of America. Whether the Jews have had to meet a religious prejudice or a racial prejudice or simply a minority prejudice, they have had to meet a hostile prejudice, and even though this may be disappearing in many cases, the Jews are well aware of it, and it has modified their psychology and their attitude toward life. When I speak of Jews and of myself as a Jew, I am merely stating the historical fact that I am descended from those belonging to a community which has had a certain tradition and body of attitudes, both religious and secular, and that I should be aware of the ways in which I myself and those around me have been conditioned by the very existence of this body of attitudes. I am saying nothing about race, for it is obvious that the Jews have sprung from a mixture of races, and in many cases are being absorbed again into another mixture. I am saying nothing about Zionism and other forms of Jewish nationalism, for the Jews are much older than any movements of this sort which have amounted to more than literary and ritual conventions, and might well continue to exist even though the new state of Israel succumbs or gives way to other manifestations of nationalism. I do not pretend to assign a normative value either to language or religion or race or nationalism, and least of all to mores. What I assert is that I myself and many of those about me come from an environment in which our knowledge of the fact of our Jewish origins is significant for our own understanding of what we are, and for our proper orientation in the world about us.

On the side of my father, Leo Wiener, the documents are scanty and perhaps largely irrecoverable. This is all the more so since the Nazi sack of the White Russian city of

Byelostok during the Second World War. There my father was born. My grandfather is said to have lost the family genealogy in the burning of a house in which he lived, although indeed from what I have heard of him he was capable of losing documents in the most calm and sober of times. As I shall point out later, there is a tradition that we are descended from Moses Maimonides, the Jewish philosopher of Cordova, and the body physician of the Vizier of King Saladin of Egypt. Even though a man is but a distant kinsman of his own ancestors of seven hundred years ago, I should like to imagine that this tradition is true, because Maimonides, the philosopher, the codifier of the Talmudic law, the physician, the man of affairs, is a much more pleasing ancestor to me than most of his contemporaries. It would scarcely be respectable to claim to be the descendant of a medieval monk—the only type of intellectual then existing in Western Christendom. But I am afraid that after so much passage of time our supposed ancestry is a very shaky legend, and perhaps is based on no more than a dash of Sephardic blood which has leaked into our veins at some epoch.

The next outstanding figure in our ancestry is a much more certain one, even though I find him far less attractive. He is Aqiba Eger, Grand Rabbi of Posen from 1815 to 1837. Like Maimonides, he was recognized as one of the greatest Talmudic authorities, but unlike Maimonides, he was opposed to secular learning, which was coming into Judaism through such men as Mendelssohn. On the whole, I feel quite content that I did not live in his times and that he does not live in mine.

My father has told me that one thread of our ancestry leads down through a family of publishers of the Jerusalem Talmud which appeared in Krotoschin in 1866. I do not know their precise relationship to my grandfather, Solomon Wiener. I only saw my grandfather once in New York when I was a small boy, and he made no particular impression on me. I gather that he was a scholarly journalist and a most irresponsible sort of person, unable to hold his family together. He was born in Krotoschin, but married and settled down in Byelostok, where my father was born in 1862. One thing he did has had a great though indirect influence on my life: he sought to replace the Yiddish of his environment by literary German. In doing so, he made certain that German should be my father's native language.

My father's mother came from a family of Jewish tanners in Byelostok. I am told that they had been honorary citizens of Russia in the old days. For a Jew this amounted to a minor patent of nobility. For example, when the Tsar came to Byelostok, it was the house of my grandmother's family that was selected as his residence. Thus, their tradition was somewhat different from the tradition of learning of my grandfather. I suspect that it was his solid, business-like habits that gave my father a firm footing in life; and although he was an enthusiast and an idealist, he had his feet well planted on the ground and was always a good custodian of the family responsibility.

Let me insert here a word or two about the Jewish family structure which is not irrelevant to the Jewish tradition of learning. At all times, the young learned man, and especially the rabbi, whether or not he had an ounce of practical judgment and was able to make a good career for himself in life, was always a match for the daughter of the rich merchant. Biologically this led to a situation in sharp contrast to that of the Christians of earlier times. The Western Christian learned man was absorbed in the church, and whether he had children or not, he was certainly not supposed to have them, and actually tended to be less fertile than the community around him. On the other hand, the Jewish scholar was very often in a position to have a large family. Thus the biological habits of the Christians tended to breed out of the race whatever hereditary qualities make for learning, whereas the biological habits of the Jew tended to breed these qualities in. To what extent this genetic difference supplemented the cultural trend for learning among the Jews is difficult to say. But there is no reason to believe that the genetic factor was negligible. I have talked this matter over with my friend, Professor J. B. S. Haldane, and he certainly is of the same opinion. Indeed, it is quite possible that in giving this opinion I am merely presenting an idea which I have borrowed from Professor Haldane.

To return to my grandmother, it is certain, as I have said, that she received very little aid from my grandfather at any time, and the young family had to be brought up to earn its own living. The age of thirteen is a rather critical one in the Jewish tradition, for it represents the admission of the boy into participation in the religious community. In general the prolongation of youth into the period of high school and college which belongs to our Western culture is foreign to Judaism. From the beginning of adolescence, the Jewish boy is given both the dignity and the responsibility of a man. My father, who was an intellectually precocious child, had begun to support himself at the age of thirteen by tutoring his fellow students. At that time, he already spoke several languages. German was the language of the family, and Russian that of the State. The role of German in his life was reinforced by the fact that because of the German bias of my grandfather my father went to a Lutheran school. He learned French as the language of educated society; and in Eastern Europe, especially in Poland, there were still those who adhered to the Renaissance tradition and used Italian as another language of polite conversation. Moreover, my father soon left the Minsk Gymnasium for that of Warsaw, where the classes were also conducted in Russian, although Polish was the language that he spoke with his playmates.

A RUSSIAN IRISHMAN IN KANSAS CITY

Father always felt very close to his Polish schoolmates. He told me that so far as he knew he was the only non-Pole at that time to be identified with the underground Polish resistance movement and to be privy to its secrets. As a Warsaw Gymnasium student he was a contemporary of Zamenhof, the inventor of Esperanto, and although the two were students in different Gymnasia, my father was one of the first to study the new artificial language.

This gives more force to his later rejection of its claims, and indeed of those of all artificial languages. He asserted, and I believe rightly, that by the time an artificial language would have developed a sufficient tradition to be used with the intellectual precision and emotional content of the existing natural languages, it would also have developed a burden of idiom structure equal to that of its competitors. Father's fundamental idea was that to a very considerable extent the difficulty of a language reflects the amount of thought that has come to make up its tradition, and that the English language is as dependent on its idioms for expressing complicated ideas as written Japanese (which can express every word by its phonetic notation) is dependent on Chinese characters for terseness. Father always considered Basic English less basic than debased. No language with idioms adequate to express complex ideas concisely, he said, would be able to serve as an easy vehicle of neutrality between competing cultures.

From the Gymnasium my father went on to the medical school of the University of Warsaw. I dare say that at least part of his motive was one so common in Jewish families, which generally desire to have at least one son a professional man and if possible a doctor. The motive is strong and easily understandable in a group which has long been undervalued in the community. Lord only knows how many tongue-tied rabbis, discontented lawyers, and physicians without a practice this motive has produced.

At any rate, my father soon found out that he had not the stomach to be a doctor. The work of dissection, as well as, I surmise, the self-defensive coarseness of his fellow students, sickened him. At any rate, he soon left Warsaw to enroll in the Polytechnicum which was then in Berlin, although it has now been situated for many years in Charlottenburg.

Father came to Berlin with an excellent secondary training. The Gymnasium, as opposed to the Realgymnasium and the Oberrealschule, stresses the classics above all else, and my father was an excellent Latin and Greek scholar. However, the Gymnasium does not scamp its mathematical training. Indeed, Father continued to be an amateur mathematician all his life, and to contribute now and then to obscure popular American mathematical journals, so that it was not until I had begun work which was already of late college or early graduate-school standard that I came to feel that I had outdistanced him.

I do not know that my father was much more successful as a budding engineer than as a budding doctor. He has told me very little of this time of his life save that it included the frugal beer, cigars, and meat cookery of the impecunious Jewish student. I do know that he worked in the drafting room between a Serb and a Greek, and that he added Serbian and modern Greek to his linguistic repertory.

My father had wealthy relatives in Berlin. They were bankers, connected with the Mendelssohn bank, with traditions going back to Moses Mendelssohn and the eighteenth century. They tried to persuade my father to join them as bankers, but he did not like the confined life, and was still hungry for adventure.

One day he happened to go to a student meeting of humanitarian nature. The speeches reinforced a vein of Tolstoyanism which had long been in him, and he decided to forswear drink, tobacco, and the eating of meat for the rest of his life. This decision certainly had important consequences for my future. In the first place, without this decision my father would never have come to the United States, he would never have met my mother, and this book would never have been written. However, assuming for the sake of argument that all these events had occurred in their due sequence, I should still not have been brought up as a vegetarian, should not have lived in a house in which I was surrounded by horrible and hair-raising vegetarian tracts concerning cruelty to animals, and should not have been subjected to the overwhelming precept and example of my father in such matters.

All this is speculation. The fact is that Father did join up with a student colleague in a wild undertaking to found a vegetarian-humanitarian-socialist community in Central America. His companion reneged, and Father found himself alone on board a ship bound for Hartlepool, after showing a bewildered official his Russian school certificate in lieu of the German military papers which he should have possessed. After crossing England to Liverpool, he sailed again for Havana and New Orleans. This was a two weeks' trip, during which my father learned the essentials both of Spanish and English. I am told that he took his English very largely out of the plays of Shakespeare. The combination of his linguistic fluency with his archaic vocabulary must have left quite a strange impression on the people he met on the levee at New Orleans. The harebrained scheme of a Central American colony had already exploded, and Father was left with his career to make in the United States.

I have before me as I write a copy of a series of articles entitled "Stray Leaves from My Life," written by my father in the spring of 1910 for the Boston *Transcript*—the dear, stodgy, civilized old *Transcript!* It is a shock for me to realize that they were written by my father when he was ten years younger than I am now. They cover his youth and education in Europe, his trip to America, and his life here until he had arrived at a successful academic career at the University of Missouri. They are written with the full romantic joy in living and the adventurous indifference to poverty and hardships that belong to a vigorous young man, more especially to one who has just escaped from the rigid discipline of a European secondary education. *Dans un grenier, qu'on est bien à vingt ans!*

The expatriate American who consciously seeks the titillations of Bohemia on the Left Bank is generally ill-prepared for this experience, and is not aware of its real significance to the young European. He has never been subject to the strict discipline which belongs alike to the French *lycée*, the German Gymnasium, and the English public school. He does not feel the utter hunger for a period of growth and freedom between his escape from this servitude and the stricter servitude of earning his own living in a hard competitive world. For him Bohemia is nothing but an extra period of laxity superimposed on the lax and undemanding education he has already known. Even worse, it is a laxity in which he has freed himself from the demands and standards of American society without assuming those of the country in which he finds himself. He is lucky if he does not give himself over completely to drink, lust, and unproductive sloth.

The European boy, on the other hand, and especially the European boy of the last century, had to burst the hard cocoon of an education, effective, severe, and traditional, and to try his wings. It mattered little whether he did this among the cloistered pleasures of Oxford or Cambridge, in the beer-and-song *Burschenleben* of the German university, or in a garret in the *Quartier Latin*. The supreme assertion of youth

A RUSSIAN IRISHMAN IN KANSAS CITY

and freedom was to wander in new lands at a time when the United States was itself emphatically a new land.

Thus Father's artless tale is written in the genuine spirit of the purely American counterpart found in Mark Twain and Bret Harte. It breathes the quintessence of youth, courage, and adventure, seen through rose-colored glasses. One senses the dust of the Southern roads and the new-turned furrow of the Kansas farm, the clamor of the raw Western city, and the keen wind off the peaks of the Sierras. All through it goes the slight, active, bespectacled figure of my father, alert to everything strange and striking, living the new life to the full, losing a job and taking a job without thought of the morrow, and having a glorious time through it all.

He was a small man, about five feet two or three inches in height, very quick in his movements, and a person who made a sharp and definite impression on everyone who saw him. He had the chest and shoulders of an athlete, with narrow hips and slender legs, and in those days he had the lean alertness of the athlete as well. His eyes were dark and flashing and beamed quick and penetrating intelligence behind the heavy glasses of the myope. His hair and mustache were black and remained black until his late middle age, and his face was ascetic. An enthusiastic walker and cyclist, he used to lead a crowd of young people on excursions into the country; and I still remember a photograph of a group of these in which he was standing beside a high-wheeled bicycle of early vintage.

He had a sharp and decisive voice and an excellent command of English, as of every other language which he spoke. I am told that he had a strong foreign accent, but through habit my ear was early blunted to this, and my impression of his English is that it showed its foreignness more in an excessive precision of diction and of vocabulary than in any other way.

He was a combative and fascinating conversationalist, although with his great intellectual power and aggressiveness it was difficult for him to limit his share in a discussion. Many times what he said would be a series of brilliant dicta rather than the give and take which best brings the other person out. He was impatient of fools, and I am afraid that to his keen intelligence very many people seemed to be fools. He was kind and beloved by his students, but he could be overwhelming through the very impact of his personality, and he was constitutionally incapable of allowing for his own forcefulness.

He was an enthusiastic farmer and outdoor man and a tireless walker. He tended to impose his amusements and preferences on those about him without fully realizing that many of them might have come to a fuller participation in a life together with him if this participation had not been so obviously enforced. One of his particular hobbies was the collection, cooking, and eating of such fungi of the region as were reasonably safe. Perhaps the infinitesimal chance of a catastrophe from eating poisonous mushrooms lent a certain savor to this sport.

He was eighteen when he arrived in New Orleans in 1880, with fifty cents in his pocket. Most of this went into his first few meals of bananas, and he had to look around for work. His first job was in a factory where cotton was baled by a hydraulic press. However, when a comrade fell into the press and was badly mangled, Father lost his interest in the work. Then he got a job as a water boy on a railroad being built across Lake Pontchartrain. He lost the job through some maladroit blunder of a boy unfamiliar with manual labor. There followed a period of aimless tramping through the deep South with a comrade or two, and after that a

A RUSSIAN IRISHMAN IN KANSAS CITY

further period of farming in Florida and in Kansas. No farmer can be more enthusiastic than the Jew when he decides to turn his hand to the plow. To my father's dying day, he was more pleased by raising a better crop than his professional farmer-neighbors than he would have been by the greatest philological discovery.

At one point in his farming career, Father ran into the remains of an old Fourierist community in Missouri. It had gone to seed, and all the efficient people had left it, while all the rogues and footless, incompetent idealists remained. Father soon had his fill of this, and while he continued to be a Tolstoyan all his life, he never afterward had much use for those whose idealism was not mixed with a certain practical sense.

I don't know just how Father came to Kansas City in the first instance, nor exactly what he did there. There was a period during which he worked as a peddler. On another occasion he pushed a broom through a Kansas City store. By this time the fun of the new American adventure was wearing rather thin. Father had begun to be a bit envious of the well-clad customers. He had decided that he was entitled to some of the pleasures and amenities of the life about him. It must have been during this time that he passed a Catholic church on which there was a sign: GAELIC LESSONS GIVEN. This was too much for Father's philological curiosity to bear. He joined the class; and because he was a far more gifted linguist than the others, he rapidly became the teacher of the class and soon the head of the local Gaelic society.

The fame of the "Russian Irishman," as he was called, was widespread about Kansas City. It had for some time been notorious in the public library that the humble immigrant peddler was calling for books that no one else could read, and was reading them.

At length my father decided to end this anomalous existence and to go back into that intellectual work for which he was cut out. He ventured to ask the Kansas City superintendent of schools for a job; and after a trial period in a wild sort of country school in Odessa, Missouri, he was taken on in the Kansas City high school. Here he showed himself a brilliant teacher, a great friend of the students, and a reformer who left his mark on the Kansas City school system. When Father was teaching (but not always when he was teaching me), he tried to draw out his students' interests rather than to compel them to think in preassigned directions. He aimed at exciting their independent thought, not their involuntary obedience. He took part in their sports and their excursions, and managed to transfer to them some of his love for the out-of-doors.

During the period of his teaching at the Kansas City high school, my father took a trip to California with some of his friends. He used to delight in telling me of the romantic city of San Francisco, of his tramp through the Yosemite Valley, and of his initiation into mountain climbing in the high Sierras. He told me that on the occasion of his climbs, he met a lady tourist, who was greatly interested in the young man's romantic love of nature and adventure. This lady was Miss Annie Peck, who later became one of the most distinguished mountaineers of her generation, and who made notable climbs in the Andes, among them the ascent of Chimborazo and Cotopaxi. Later on Miss Peck wrote to Father, saying that her alpinism was first stimulated by his enthusiasm.

One of my father's amusements during his Kansas City period was to attend spiritualistic seances and to try to discover the sleight-of-hand technique of the mediums. I don't think that Father was very much excited with the ideas of spiritualism either pro or con, but the chance to do a little detective work appealed to his adventurous spirit and to his intellectual curiosity. He became firmly convinced that if there were anything whatever in spiritualism, it was not to be found among the mediums he had investigated.

The budding culture of the Middle West of this period was attracted to the anfractuous style and the puzzling allusions of Browning's poetry. Of course, to a man of my father's breadth of cultural background neither the style nor the allusions offered much difficulty. Father became rather a lion to roar at the meetings of ladies' Browning clubs, and I believe it was there that he met my mother. Be that as it may, they certainly enjoyed reading together *The Ring and the Book* and *On a Balcony*. My own name and that of my sister Constance represent characters from *On a Balcony,* and we are thus unwilling fossils of a bygone intellectual era. I must think that my parents' indifference to the consequences of giving me such a recondite and unusual name was part and parcel of the decision which they had already made to direct and to channel my life in every detail.

My father chose to become a teacher of languages. He might almost as readily have become a teacher of mathematics, for he had both talent and interest in the field. Indeed, throughout my college training I learned the large part of my mathematics from him. There are times when I think that it would have been more fortunate for my father if he had taken mathematics for his field rather than philology. The advantage is that mathematics is a field in which one's blunders tend to show very clearly and can be corrected or erased with a stroke of the pencil. It is a field which has often been compared with chess, but differs from the latter in that it is only one's best moments that count and not one's worst. A single inattention may lose a chess game, whereas a single successful approach to a problem, among many which have been relegated to the wastebasket, will make a mathematician's reputation.

Now, philology is definitely a field dependent on the careful assessment of a number of small considerations rather than on a mechanical ride in a train of logic. For a man with intuitions and imagination, philology is a field in which he may easily go wrong and in which, if he does go wrong, he may never find it out. The mathematician who makes serious mistakes and never finds them out is no mathematician, but an imaginative philologist may go very far wrong before a demonstrable error pulls him up sharp. My father's success in philology was unquestioned, but his sanguine temperament would have benefited under the discipline of a field in which discipline is automatic.

This, then, was the strange young man who became my father and my teacher. In 1893 he married my mother, who had been Miss Bertha Kahn, the daughter of Henry Kahn, a department store owner, of St. Joseph, Missouri. Let me say something of my mother and her background.

« I I »

THE PROPER MISSOURIANS

HENRY KAHN, my mother's father, was a German Jewish immigrant from the Rhineland, and a department store owner of St. Joseph, Missouri. His wife belonged to a family named Ellinger, which had been settled in the United States for at least two generations previously. I gather that my grandmother's mother was not Jewish. This seems to have introduced a peculiar pattern of marriages into the Ellinger family, so that the girls of that generation tended to marry Jews like their father, and the boys Gentiles like their mother. At any rate, even a hundred years ago the family was hovering in a state of unstable equilibrium between its Jewish background and an absorption into the general community.

This tendency for the marriage pattern of the boys to differ from that of the girls is something of which I have heard under quite different circumstances. I have been told of a family in New York in which the wife was Dutch and the husband a Chinese Protestant clergyman. All the boys of that marriage seem to have lost themselves in the general American background, taking American wives and putting behind them the Oriental part of their ancestry. The girls, on the other hand, all married Chinese and returned to China. In that case, the motive for the differentiation seems to have been the extreme demand on the part of young Chinese men for wives with a Western education and background, which gave the daughters unusually favorable opportunities for Chinese marriages as against American marriages, conditioned at least in part by the restrictions of race prejudice. Whether motives paralleling these to at least some extent existed as well in the Ellinger family, I do not know; but it is interesting to observe the phenomenon cropping up in such different situations. The Sword follows the Sword, the Distaff the Distaff.

The Ellinger family appears to have had its American origins in Missouri and perhaps even further to the south. Its members combined a proper Southern gentility of outlook with a high degree of unpredictableness. More than one of the men ended a career of impeccable propriety by suddenly leaving his family and taking to the great open spaces. There is a legend that one of the Ellinger family finally became a Western bandit and was shot down while resisting capture.

Even apart from such radical manifestations of individuality, the Ellingers were and are a centrifugal clan. It is only in the later history of the family that the gradual dilution of these idiosyncrasies has permitted them to take the full place in the community that their abilities have always justified.

We are now used to the fact that almost all of us are the descendants of immigrants. This was not true in the middle of the last century. Today the melting pot not merely melts but alloys. It does this much more easily because it is not overloaded with strange and refractory metals which have been thrown into it cold. The immigrant family which has already begun to lose itself in the general American picture is no longer faced by the even more recent immigrant who has just stepped out of the steerage. No longer is our American of Continental ancestry confronted with a hierarchy of greenhorns, established immigrants, and old Americans, forming a stable ladder of social ascent in which each man has his rigid place.

In many ways the early immigrants who had the easiest emotional time of it were the Eastern European serfs, who had almost literally nothing to lose but their chains. For the immigrants of higher caste in the European system, their Americanization and subsequent ascent in the social hierarchy was preceded by a disinheritance and a leveling downward.

All this was inevitable and perhaps even an essential part of the discipline which the foreigner needed to fit him to take a place in a community very different from the one into which he was born. Today, however, the immigrant is not only the beneficiary but a benefactor to the country to which he has immigrated. His native culture often has a richness which should not be degraded and lost in the general mist of a tradition averaged thinly over a continent. There are facets of his art and his thought, of his folklore and his music, which are well worth resetting in the regalia of America. Yet in the presence of the overwhelming invasion of greenhorns from the old country, these heirlooms of the immigrant were difficult to evaluate and to appreciate. He was only too likely to accept without protest the inferior position which was allotted to him, and to protect his ego by a similar depreciation of the even more recent immigrant.

In such a community and such a period, respectability is a pearl beyond price. Schmidt becomes Smith, and Israel Levin becomes Irving Le Vine. The evangelical (and, incidentally, also the rabbinical) religious injunction to avoid even the *appearance* of evil is interpreted as an injunction to avoid the appearance of evil and vulgarity far more than evil and vulgarity themselves. A strong character may indeed defy such a society and live according to his own values. It is much easier for the less forceful character to accept these values, and to fall on his knees before Mrs. Grundy. Only a man like my father, who was ready to defy the almighty Jehovah himself, was not likely to fall into the orthodoxy of Grundy worship.

The specific frustrations of the immigrant and Jewish families of the seventies and eighties and nineties were reinforced by the general moral backwater of the Gilded Age. It was an age in which the Grant of the Whiskey Ring had replaced the Grant of the Civil War and in which Lincoln was dead and the Daniel Drews and the Commodore Vanderbilts were very much alive. The enthusiasms and devotions of Civil War times had run out and the enthusiasms and devotions of the twentieth century had not yet appeared above the horizon. There was a general slackness and letdown. This letdown must have been most intense in the defeated South, and in the Missouri which was almost an anteroom to the South.

Besides these stresses and strains belonging to the social group and to the time, my mother's family was also subject to more personal dissensions. Her family was split by an alienation between her parents. My mother's mother was a person of good general culture and vigorous and uninhibited emotions. She had a strong, undirected vitality which enabled her to live to a great age, and she was simply too much for her quieter and less energetic husband.

Then, too, one of my mother's older sisters had ambitions to be the woman intellectual of the family, and rather looked down on her sisters. This led to an ultimate rift, in which my mother and father stood on one side and most of her family on the other. At least one cause of this rift was the traditional friction between the German Jew and the Russian Jew and their differences in social status. This was supplemented by my father's downrightness and naïveté in social matters.

In any case, my mother, with my father's support, gradually broke with her family. Although he must often have been a puzzle to her, she was deeply in love with my father, and admired him greatly. Nevertheless, it was not an easy step for my mother to take. She had been brought up with the indulgence often extended to the belle of the family. I remember one photograph of her made when I was about four years old. She looked extremely handsome in the short sealskin jacket of that period. I had great pride in that picture and in her beauty. She was a small woman, healthy, vigorous, and vivacious, as she has indeed remained to this present day. She still carries herself like a woman in her prime.

In the family of divided roots and Southern gentility into which she was born, etiquette played a perhaps disproportionately large part, and trespassed on much of the ground which might be claimed by principle. It is small wonder, then, that my mother had, and conceived that she had, a very heavy task in reducing my brilliant and absent-minded father, with his enthusiasms and his hot temper, to an acceptable measure of social conformity.

The permissions and demands of our society in this matter differ widely for the man and the woman. For the man, a certain degree of shaggy unconformity may be permitted in return for character and genius. But the woman is expected to be the custodian of the more conformist and conventional virtues, which indeed need someone to cultivate them. A man may have a hot temper without reproach, but a woman must be smooth and suave. When I was born there was added to my father's downrightness the problem of a child with somewhat similar ability, the same hot temper, and the same resistance to taming; and it is not surprising that my mother sometimes must have felt at her wits' end. Later, when my father's temper and mine came into head-on collision, mother could do little but act as a general peacemaker, without indulging in any too definite opinions or convictions of her own on which to base this peacemaking. This made it hard for me to understand her. In my collisions with my father, dramatic as they were, I could generally recognize a principle which I had to respect, even when I was suffering from my father's interpretation of it. My mother was scarcely able to afford such luxuries. When the husband is a zealot, the wife must be a conformist. How many unworldly scholars, whether Jews or Christians, must have depended for their very existence on their conformist wives!

When my parents married, my father was already Professor of Modern Languages at the University of Missouri in the town of Columbia. He taught both French and German, and my parents shared in the simple social life of a small college town. They lived in a boardinghouse together with many members of the faculty, and there I was born on November 26, 1894.

Of course, I have no memories of the town which I left as a babe in arms, but there were family stories which go back to that boardinghouse, and to Father's friend of those days, W. Benjamin Smith (who later taught mathematics at Tulane University). He was a particular crony of my father's and a great practical joker. One time Smith came back to the boardinghouse and found that the colored waiter of ample dimensions had been replaced by a wizened little fellow. "Sam!" roared Professor Smith, "how you have shrunk!" The puzzled waiter left the room on the run, never to return.

I mention a story concerning Smith and a Negro since it was on this racial issue that his friendship with my father broke up many years later. Smith, who was an unreconstructed rebel, had published a pseudolearned book on the inferiority of the Negro, and that was too much for my father's liberalism, and for his respect for facts.

As little more than an infant, I accompanied my parents on a one-way trip that ended in Boston. The motive power behind this trip was a power rooted deeply in Missouri. A Missouri politician had his eye on my father's job for kinsman or henchman. My father had met with such success that it was no longer possible to run Modern Languages at the University of Missouri as a one-man show. It was decided to put a man in charge of each of the departments of French and German. When offered his choice, my father chose German. Unfortunately, there was a protégé or connection of the politician's family who also had his eye on this chair, and Father was left out when the department was split. Father had no connections in the academic life of the rest of the country. He came to Boston on a pure speculation, for he thought that it was best to look for jobs where the jobs were.

He soon attracted the attention of Professor Francis Child, the learned editor of *Scottish Ballads*. Child traced the Scottish ballads and their parallels through the various languages of Europe and Asia, and needed help in collating sources for many of them. Father was given the southern Slavic languages as his assignment. He made himself so useful to Child that Child helped him find a position around Boston. Father's first teaching jobs were at Boston University and the New England Conservatory of Music, and he did some work in the cataloguing department of the Boston Public Library. Finally, Child obtained for him an instructorship in Slavic Languages at Harvard, the first of its kind at Harvard and, I believe, in the country. This led to a career of gradual promotion through the successive ranks of assistant professor and professor until my father retired in 1930.

Nevertheless, for many years he had to reinforce his salary with outside jobs. Although living expenses were low, salaries were also low. Father continued his work at the New England Conservatory and at Boston University for several years, and did occasional jobs for the Boston Public Library. He also did a considerable amount of important etymological work for several editions of the Merriam-Webster *Dictionary*; in this work he was associated with Professor Schofield, also of Harvard. In later years, Father's main source of academic pin money was Radcliffe College, which had served Harvard professors for so many years as a supplement to salary.

Professor Child was a remarkable and most democratic person, and a sincere friend of my father. One day Father saw another short, nearsighted, vigorous young man leaving Child's house. When Father came in, Child told him that he had just missed meeting Rudyard Kipling. Apparently Kipling had got off to a bad start with Child, who had been out in his garden in his old clothes watering his roses. Mr. Kipling had mistaken him for the hired gardener. "Ah," said Child, "a workman looked over my fence yesterday and was drinking in the odor of my flowers. *He* was my brother."

« I I I »

FIRST REMEMBERED PATTERNS

1894–1901

IT IS A DISSERVICE of some Freudians (and I do not mean Freud himself) to have reduced the infant to a homunculus possessing little mental life outside of a rudimentary sexuality. Many a Freudian looks askance at all other recollections from infancy and very early childhood. I will not for a moment deny that infantile sexuality exists and is important. But it is far from an exhaustive description of the child's early mental life, both emotional and intellectual.

My conscious memory goes back to a time when I was about two years old, when we lived in a second-story apartment on Leonard Avenue, in a rather obscure and not too desirable region on the boundaries of Cambridge and Somerville. I remember the staircase leading to our quarters: that it seemed to rise for what was for me an interminable distance. We must have had a nursemaid even at that remote period, for I recollect going out with her to make purchases at one of the little shops which I was told was in Somerville. The whole region is a confusion of streets belonging to the unconformable systems of the two cities, and I distinctly remember the acute angle at which these streets intersected before our particular grocery shop.

Around the corner was a rather grim and terrifying building which, I learned, was a hospital for incurables. It still stands, and is the Hospital of the Holy Ghost. I am quite certain that I had at that time no clear idea of what a hospital was, and even less idea of an incurable, but the tone in which my mother or our nursemaid mentioned the place was enough to fill me with gloom and foreboding.

This is all that I can truly remember about Leonard Avenue. I was later told that my mother had a second child there who died on the day of its birth. When I was told this, I was a child of thirteen; the news shocked me to the extreme, for I was afraid of death and had comfortingly believed that our own family circle had never yet been broken. I have no direct recollection of the existence of this child, and I still do not know whether it was a boy or a girl.

We spent the summer of 1897, when I was two and a half years old, in a hotel in Jaffrey, New Hampshire. There was a pond nearby with rowboats on it, and a path leading up a mountain whose name I understood to be Monadnock. My parents climbed the mountain, naturally without me, and they also took me to a neighboring village where for some reason or other they visited the blacksmith shop. The blacksmith had had a toe crushed from a horse stepping on it, and I was frightened to hear of it, for even at that time I had a very lively terror of injury and mutilation.

The academic year of 1897–98 found us on Hilliard Street in Cambridge. I have a dim recollection of seeing the moving van which transferred our goods from Leonard Avenue. From this point on, my recollections come thick and fast. I remember

FIRST REMEMBERED PATTERNS

my third birthday, and my two playmates, Hermann Howard and Dora Kittredge, the children of Harvard professors who lived on the same street. I am sorry to say that my first recollection of Hermann is quarreling with him at his own birthday party, when he was five and I was three.

I am told by my parents that while we lived on Hilliard Street I was taught French by Josephine, a French maid who worked for us. I have no recollection of Josephine herself, but I do remember the children's textbook which she used, with the names and pictures of a spoon, a fork, a knife, and a napkin ring. What French I learned at this time I must have unlearned with equal rapidity, for when I studied French again in college at the age of twelve, no obvious trace of my former knowledge of the language remained.

It must have been Josephine who took me for walks on Brattle Street and about the Radcliffe grounds. The darkness of what I now feel as the pleasant shade of the trees on Brattle Street terrified me at the time; the geography of the neighboring streets left me in utter confusion. There was a house on the corner of Hilliard and Brattle Streets which had a closed-off bay window that frightened me very much because it looked as if it were a blind eye. I had the same feeling of terror and claustrophobia when my parents had the carpenter close up some sort of serving hatch connecting the dining room of our house with the butler's pantry.

Not far from our house was an old school building, but whether it was used or deserted at the time I do not remember. Mount Auburn Street was only a few houses away, and around the corner was a blacksmith shop, with a driveway lined with white painted cobblestones. I tried to lift one of them once and take it away, and I was duly reproved. An alley by the side of our house led back to a little garden where an old gentleman by the name of Mr. Rose—at least he seemed old to me—would take the air and smoke his pipe. Behind that was another house where there lived two older boys who took me under their wing. I remember that they were Catholics and had in their house an image of the crucified Christ with the wounds and a crown of thorns which struck me as the image of a victim of cruelty and injustice. They also had a potted plant which they called Wandering Jew, and to explain its name they told me a legend which I did not understand but found very painful.

Over all this early period of my life, I have very little recollection of my father. My mother figures largely in my early memories, but my father was an austere and aloof figure whom I saw only occasionally in his library, working at his great desk. I used to play under that desk. I have no recollection of any coldness or harshness on his part, but the low timbre of the male voice was in itself enough to scare me. To the very young child, the only parent is the mother, with her solicitude and tenderness.

Mother used to read to me in the garden. I know now that the yard was a mere three or four feet of grass outside the front steps, but then it seemed to me enormous. The book from which she most enjoyed reading was Kipling's *Jungle Book*, and her favorite story was "Rikki-Tikki-Tavi." I myself was beginning to read at the time, but I was only three and a half years old, and there were many words that caused me difficulty. My books were not particularly adapted to my years. My father had an old friend, a lawyer by the name of Hall, who was blind in one eye and deaf in one ear, and quite abstracted from human society and ignorant of the needs of a small child. He gave me for my birthday the volume on mammals of the old *Wood's Natural History*. It was in small type, and an nth reprint at that, with the type and the woodcuts blurred and clogged with ink. My parents lost the original copy, but in order not to disappoint the old gentleman they

FIRST REMEMBERED PATTERNS

promptly got another, and even before I could read it with any ease, I used to finger through the pictures.

Another book which I received as a present about the same time is rather a puzzle to me. I know that it was a children's book on elementary science, and I know that among other things it discussed the planetary system and the nature of light. I know that it was a translation from French, and that at least some of the woodcuts represented Paris. However, I do not know the name of the book, and I do not believe that I have seen it since I was five years old. Perhaps it was a translation of a work by Camille Flammarion. If any of my readers can identify the work and I can lay my hand upon it, I can certainly check up from my memory of the pictures whether it is the book to which I am referring. As I have made my career in science, and as the book was my first introduction to science, I should very much like to see the point from which I started.

I cannot remember many of my toys of that time. One, however, which I can remember with absolute clarity, was a little model of a battleship which I pulled along on a string. It was the time of the Spanish-American War, and toy battleships were all the rage among the youngsters. Even now I can recall the white paint and the straight masts of the battleship of that transitional period before the days of the dreadnaught, with the small deck turrets of the secondary battery, and only a few turrets containing guns of the larger caliber.

My nursery was a room at the back of the house, separated from other rooms on the second floor by a step or two. One day I stumbled and fell up this little flight of steps, gashing my chin and leaving a scar which I bear to this present day and which is one of the reasons for my wearing a beard. I also cut my hands on the metal fins of the little iron cot in which I slept. I can still recall just how uncomfortable it felt.

I remember the songs with which my parents sang me to sleep. My mother was a great lover of *The Mikado,* and its arias are among my very early recollections. Some music-hall songs also played a role in my childhood, and among these were "Ta-ra-ra-boom-de-ay," and "Hush, Hush, Hush! Here Comes the Bogey Man." My father preferred the "Lorelei," and a Russian revolutionary song which I never understood but whose syllables I remember to the present day.

My sister Constance was born in the early spring of 1898. The midwife, a genial Irishwoman by the name of Rose Duffy, was a particular friend of mine, and I named a rag doll after her. She lived on Concord Avenue with a sister, Miss Mary Duffy, who did her housekeeping. When I visited them I had the run of their box of gingersnaps and molasses cookies.

I am told that the arrival of my sister disconcerted me very much. Certainly some years after, when she was old enough to be an individual, I did begin to quarrel with her in a most reprehensible way, but this was succeeded by many years of companionship and good feeling. To have a baby in the house taught me much I did not forget later about the mysteries of bottles and diapers.

During that summer, Father traveled in Europe. I was delighted with his post cards from strange cities, with the text printed out for me in consideration of my childish inability to read handwriting. Also during that summer I began to read a certain natural history magazine which had pictures of birds. I can even recall the queer, old-fashioned advertisements on the pages of this magazine, but its name has escaped me.

Already Father had much contact with the staff of the Boston Public Library. One of his friends there, a Mr. Lee, had a wife who was an illustrator and author of children's books, and a little daughter of my own age. They lived in Jamaica Plain, within a stone's throw of Franklin Park. I remember reading Mrs. Lee's books and playing with the little girl in the stone grottoes of that part of the park. I remember

FIRST REMEMBERED PATTERNS

the streetcar trip by way of Central Square and the Cottage Farm Bridge, across a part of Boston which has completely changed character since that time. I used to read the Lee girl's copy of *The Arabian Nights*. Some years later she fell victim to diabetes, which was a death sentence to a youngster in those days before insulin. Mr. Lee gave me the book, together with some other belongings of his daughter, but it was a sad pleasure for me to read it.

One of the other books which I read at the time was *Alice in Wonderland,* but it took me years to get the full flavor of Lewis Carroll's humor, and the metamorphosis of Alice in the book had a sort of terror for me. Indeed, when I saw a copy of *Through the Looking Glass* I lost all sense of humor and flatly regarded the book as superstitious.

I was an easily frightened child. One time when my parents took me to the old Keith's vaudeville theater for want of a baby-sitter, I saw a pair of slapstick comedians hitting one another about. After a sudden blow, one of them appeared in a glaring red wig, and the whole thing scared me so much that I burst out weeping and had to be taken out of the theater.

The next year after Father had returned from Europe, we continued to live in the same house on Hilliard Street. I was sent to a kindergarten on Concord Avenue, opposite the Harvard Observatory. I have not forgotten the woolies and long leggings which I put on, or the conducted games with the other children, or the webs of paper which we had to weave. I met my first sweetheart there, a dear little girl whose voice charmed me, and in whose neighborhood it was good to be. I recall the delightful visit we kindergarten children made to a nearby garden with its crocuses and tulips and lilies of the valley under the broad spreading fir trees.

We spent the summer of 1899 at Alexandria, New Hampshire. At four and a half, I was old enough to look out of the train window and to watch the landscape flash backward. I had already begun to be interested in the technique of the railroad. By this time I must have had a toy "puff-puff" of my own to add to my interest.

From that time to about 1933 I had no occasion to revisit Alexandria. When I first revisited it I found that my memories of the geography of the place exactly fitted the scene before me: Bristol, with its Civil War monument and its old-fashioned mortar in the center of the village square; Newfound Lake, the boardinghouse where we stayed, the little house opposite it where a colleague of my father's had lived and with whose son I had played. All were just as I had pictured them from memory. I found the village of Alexandria itself unchanged, and Bear Hill, where I had taken a walk with my parents through the pine woods with their Indian pipes, and had had to be carried back on my father's shoulder. It all was just as I had fancied it. I well remember the textile mill at Bristol, with its whirring looms, where my father had taken me as a boy.

We spent the next winter—that of 1899 to 1900—in one half of a double house on Oxford Street in Cambridge. My parents already had plans to put me in school, and they took me with them to visit Miss Baldwin, the principal of the Agassiz School, which was only two doors from us. No final arrangement was made to put me in school. Miss Baldwin, an extremely distinguished schoolteacher and a woman of great dignity, was a Negro. She had entered the Cambridge school system in the 1880's at a time when the humanitarian impulse of the abolitionists had not entirely died out and before New England snobbery had surrendered to Southern gentility at the beginning of the twentieth century.

I received the *St. Nicholas Magazine* as a birthday present while I was living on Oxford Street. I remember very well the day on which the postman brought me one back number and one current number dated in 1899, and from then on, the new

FIRST REMEMBERED PATTERNS

century had begun and I was in 1900. *St. Nicholas* was a revelation to me and constituted much of my most pleasant reading in my childhood. It is scarcely possible for me to conceive how the present generation of children gets along without it or its equivalent. *St. Nicholas* always assumed that the child, for all his few years, was an essentially civilized individual, and disdained to put before him intellectual pabulum which in essence would not be worthy of an adult. How the present generation of children has been able to substitute for this the blatant and inane fare of the funnies on the one hand and the highly artistic picture book without literary content on the other is a profound mystery to me. The children of my day would consider that the children of the present day are letting themselves be shortchanged.

The autumn marked the triumphant return of Admiral Dewey to Boston after the Spanish-American War. My parents took me to see the parade that marked this occasion; but I did not and could not have had a sense of the historic importance of the occasion, for a war meant to me certain military toys that went bang-bang rather than anything that had to do with real people losing real lives.

Another clear recollection of this winter is of Christmas. I woke up in the morning well before dawn to look into my stocking and to discover what Santa Claus had put in it. At this time I did not know that Santa Claus was my father, but I appreciated the sweets and toys and little written jests that I found with the tangerine and the nuts and candy in my stocking. The larger presents were under the tree, and my sister and I should have waited until morning to see them, but we interpreted morning in a very loose sense, and made our way downstairs around four o'clock.

The other pictures I carry of the year are completely isolated. Our next-door neighbor, a militant Irishwoman, the wife of a policeman, valiantly drove off from her fortress some naughty boys who were invading it. I believe she used a broom. I rode up and down the sidewalk on my tricycle, and used to meet on my daily rides boring grown-up friends of my father. There was a one-legged boy in the neighborhood who puzzled me by appearing alternately with and without an artificial leg; he used to pass our house on his way to school on a bicycle.

It is strange how pictures of suffering and mutilation recur among my very early impressions. I doubt whether much of my interest in these matters was humanitarian or proceeded from a genuine compassion for the suffering. Part of my interest was the cruel, staring curiosity of the child, and another part was the genuine fear of catastrophe as something which had occurred to people whom I could see and which might conceivably occur to me. I had had a minor surgical operation for tonsils and adenoids about this time and was terrified by the harsh swirl of confusion that followed the administration of the anesthetic. But I perceived no relation between my own minor surgical operation and my terror at the fact of mutilation. All this is a repetition of an observation amply familiar to the Freudians, and efficiently explained by them.

My father was very proud of his early record as a farmer and had long aspired to be a landed proprietor. In this there was a mixture of Tolstoyanism and a pride in having overcome one of the traditional limitations of the Jew. In the spring of 1900 he bought the farm of his dreams in Foxboro. The house was set well back from a road, lined with a row of catalpa trees which gave the place its name: Catalpa Farm.

I do not remember what Father raised on the place, although I cannot imagine him not farming. I am sure that that summer contributed greatly to my knowledge of country life and of the New England trees and plants. The country children on the next farm took the usual advantage of me which they considered themselves entitled to take of a city boy, and filled

FIRST REMEMBERED PATTERNS

my mouth with the filth of the road. I found a few more suitable playmates in the village, or at least a few playmates more willing to accept me. They acquainted me with the existence of earthworms, and surprised me by exhibiting the very moderate inconvenience which the worm suffers by being cut in two. The cruelty of this process gave me only some minor twinges of conscience.

I remember very little of Foxboro as it was at that time, although I am sure that the most exciting thing about it for us was the gossip concerning a church which had recently been founded there by the Holy Rollers. I also remember an older boy who took me to a baseball game between Foxboro and Attleboro. The mysteries of baseball were too much for me, and it was only much later that I was able to take an interest in the game.

The early summer was marked by the arrival of my Grandmother Wiener from New York, accompanied by my cousin Olga. I remember Grandmother even then as an old lady, although she cannot have been much older than I am at the present time. She was always dressed in the dark clothes of an elderly European, and she had definitely foreign mannerisms, exhibited in the way she would gesture with her forefinger and would shake her head. She was a tiny, active person, with the air of having suffered much; and from everything that I have heard about my grandfather, he was not a man with whom it was possible to live without suffering, if only because of his temper and improvidence. In Europe Grandmother had been thrown on her own devices for earning her living; and now that my father's example had been followed by the other Wieners and they had come to America too, she was passed on from child to child according to their financial ability to support her.

Grandmother always spoke with a strong accent and was never able to distinguish between the words "kitchen" and "kitten." She read her own newspaper, printed in a foreign type-face which I later found out to be Yiddish. She always signalled her coming to visit us by bringing us dainties and toys from New York, but we would have loved her even without this. My mother, who looked rather askance on my father's New York kinsfolk, from the vantage point of having been in the land for one more generation, could not help loving Grandmother, or *Grossmutter*, as we always used to call her to distinguish her from Grandmother Kahn.

Cousin Olga was a sharp young person of nine years, four years older than I. Her mother, my Aunt Charlotte, had been left alone in the world by a runaway husband, and it was very important for Olga to have a chance for a healthy summer vacation in the country. She and my mother were always at swords' points. The life of the New York back streets tends to make one worldly-wise before one's time and this was hard on my mother.

Olga and I often quarreled. On one occasion, she and I got into a quarrel—I do not know about what. Olga told me that God knew everything, and would not approve of my behavior. I then and there made the statement that I did not believe in God. Seeing no lightning descending from heaven to smite me dead on the spot, I persisted in my godlessness and told my parents about it. I found enough sympathy in my father's attitude to encourage me to persist in this point of view.

I have never made up my childish quarrel with Jehovah, and a skeptic I have remained to the present day, although I look askance at those skeptics who make their skepticism into a positive religion, and are Bezbozhnik in the same spirit in which they might be churchmen.

There were lilac bushes under the catalpa trees, and in them I found a little nest with blue eggs. Olga told me that because I had touched them, the mother bird would leave them alone and never come again, and that the eggs would not hatch or the nestlings would die. For a child of five, that made me as good as a murderer in my own mind, and the sense of guilt that it gave troubled me for a long time afterward.

Father took me on several excursions about the neighborhood, some of them for the sake of the tramp and the chance to engage in his favorite sport of collecting mushrooms, and some of them to contribute to my education. For example, he took me to a foundry and machine shop in the neighborhood. The blast furnace was fed with scrap, not ore; and I saw the metal run into the molds for pigs and the more elaborately shaped molds for parts to be machined. The machine shop worked in brass as well as in steel. It was a delight to see the white and yellow shavings turning up into curls under the pressure of the tool.

Father often tried to secure my education at school, but he found some obstacles in his way, though I don't quite know what the difficulties were. I fancy I was just too young for the standards of the school board. I was vaccinated for the village school, and went to it for a few days, and then Father transferred me to a little red schoolhouse in the country, where children of all ages studied together under one teacher. All that I remember of it is that there was a pond outside the schoolhouse, and that it was winter, and that children were sliding and skating on it.

Some time in the spring of 1901, when I was six years old, we took rooms in a boardinghouse on Concord Avenue opposite the Harvard Observatory in Cambridge. We had returned to a Cambridge boardinghouse because we were contemplating a summer trip to Europe. My parents were busy making the necessary purchases of equipment for the trip, and buying toys and other amusements to occupy my sister and me on the ship. I have not many recollections of this time except that I again visited my playmate Hermann Howard, and that an older girl, Renée Metivier, who was lodging in the same boardinghouse, took me under her wing. She taught me how to make and to fly a kite, and I remember going down with her to Church Street to get the material. Church Street at that time was even more the artisan's section of Cambridge than it is at the present day. While I was in kindergarten the teacher had taken us down there to see the delightful mysteries of its forges and wheelwright shops and carpenters' yards.

There is one point which I would append to the discussion of my early reminiscences. It is probably of considerable interest to the reader to know how the very early intellectual development of the prodigy differs from that of other children. It is, however, impossible for the child, whether he be prodigy or not, to compare the earlier stages of his intellectual development with those of other children until he has reached a level of social consciousness which does not begin until late childhood. To say that one is a prodigy is not a statement which concerns the child in question alone. It is a statement which concerns the relative rate of his intellectual development with that of others. And it is a thing which his parents and teachers can observe far earlier than he can himself. In one's earlier stages of learning, one is one's own norm, and if one is confused, the only possible answer is that of the Indian, "Me not lost, wigwam lost."

I was well along in childhood, probably seven or eight, before I knew enough about the intellectual development of other children to comment in my own mind on their relative speed of learning and my own. By this time the earlier stages of the process of learning how to read and even the simpler aspects of learning my arithmetic had receded into the past almost as thoroughly as the average child's consciousness of learning how to speak. For this reason what I shall have to say about these matters will scarcely be distinguishable from the history of any other child, except on the basis of the precise

FIRST REMEMBERED PATTERNS

year and month in my life at which I passed various stages of development.

For mark this well: all early learning is a miracle, even on the part of the child whom we later consider to be somewhat dull. When a child begins to speak, it has already learned its first foreign language. Between birth and the age of two years there is a blossoming out of new intellectual acquisitions which can never be paralleled in later life, and this whether the child is a genius or a moron. This is a development of doing rather than of reflection about doing; a spontaneous burgeoning of new talents and not the work of the child as his own self-conscious schoolmaster. It is a fact that in my case the beginnings of reading go back to an age not twice that of the beginnings of speech with many children, and that this is obscured by the fact that I was learning to read, not learning to think about reading. Later, when I went through my earlier schoolbooks (at home under my parents' guidance), I learned some of the puzzling distinctions between capitals and small letters and script. I have memories only of the obstacles in my way, and not of the greater part of the task, which was accomplished spontaneously and unconsciously. I remember that the similarity between i and j puzzled me, and that in old-fashioned books there was a long s which looked curiously like an f. I remember the mechanical difficulty of writing, and that my best handwriting was and long remained below the acceptable standard of the class. As to arithmetic, I counted on my fingers and continued that long after it was regarded as unpermissible by the standards of my school classes. I was puzzled by such things as the axiom that a times b equals b times a, and I tried to clear this up by drawing a rectangle of points and turning them through a right angle. I was not particularly fast in learning my multiplication tables or, in fact, anything else that had to be learned by rote, although I had a good understanding of the principles of fairly complicated operations from a very early period in my early childhood. I remember the old Wentworth *Arithmetic*, in which I read ahead into the discussion of fractions and decimals without any great difficulty. In general the two things that held me up were at the opposite ends of the game: the technique of adding and multiplying rapidly and precisely and the understanding of why the various laws of arithmetic, the commutative and the associative and the distributive, were true. On the one hand, my understanding of the subject was too fast for my manipulation, and on the other hand, my demands in the nature of fundamentals went too far for the explanations of a book devoted to manipulation. But if we go beyond that to the very first beginnings of my arithmetic, they are nearly as hard for me to recall as the beginnings of my reading or speaking.

This relegation of the difficult and the truly intellectual part of my work to a level below full consciousness is not merely a matter of my childhood, but something that has continued to the present day. I do not fully know how I get new ideas or how I resolve the apparent contradictions between those already in my mind. I do know that when I think, my ideas are my masters rather than my servants, and that if they resolve themselves at all into a usable and understandable pattern, they do that at a level of consciousness so low that much of it happens in my sleep. I shall have to speak of this elsewhere, but I cannot find in my own intellectual history any brusque change between the striving of childhood after childish knowledge and the power and the striving of my grown life after the new and the unknown. I know more and I have better tools, but it would often be hard for me to say just when and how I have acquired these tools and this new knowledge.

One thing that I share with my father is an excellent memory. By this I do not mean that we cannot and have not been perfect examples of the absent-minded professor, and

FIRST REMEMBERED PATTERNS

that our capacity to forget in matters of daily life has not been ample. But I mean that when we have acquired a range of ideas or a way of looking at things, this has become a part of us not to be lost through any vicissitudes whatever. I remember the last stages of my father's life when he was on his death bed with apoplexy, and when his fine intelligence no longer enabled him even to recognize the loved ones about him. And I remember that he spoke as though he had the gift of tongues, in English, in German, in French, in Russian, in Spanish. Confused as to what he saw about him, his languages nevertheless were clear, grammatical, and idiomatic. The pattern went through the fabric, and neither wear nor attrition could efface it.

« I V »

CAMBRIDGE TO CAMBRIDGE, VIA NEW YORK AND VIENNA

June–September, 1901

G. K. CHESTERTON has said somewhere that the best way to see London is to make a voyage from London to London around the world. We cannot appreciate an experience until and unless we have other experiences which differ from it to serve us as points of orientation. I am certain that I could never have learned to understand New England if I had not at some period in my life got far enough away from it to see the great lines of its spiritual character spread before me as a map.

In the late spring of 1901, when I was six and a half and my sister was about three, our family took the Fall River boat to New York where we were to stay with my father's relatives. They were living somewhere in the East Sixties between Third and Fourth Avenues. Upper Fourth Avenue had not at that time gained its present prestige as Park Avenue; the region was a slightly superior East Side slum. The typical old-fashioned apartment in which my relatives lived was at the top of a long flight of outside steps leading in from the street.

CAMBRIDGE TO CAMBRIDGE, VIA NEW YORK AND VIENNA

It was dark, overcrowded, and stuffy. The only windows were at the very front and at the very back, and the apartment was already more than full before it had to take the burden of four more visitors from Boston. But we were near enough to Central Park to be able to reach it by a short walk past the palaces then lining Fifth Avenue, and the Central Park Zoo was always a delightful goal for us.

My Uncle Jake Wiener was the only man of the establishment. He was a journeyman job printer and very good at his trade. His main amusement was gymnastics. At one time he had been number three man in the American rating on the parallel bars. Certainly, if ever a man was built for the sport it was Uncle Jake. I have said that my father had powerful shoulders, but his brother Jake's were enormous, and he was muscled like a wrestler. He was even shorter than my father, who was also a short man; and his legs were thin and spindling. He had the drawn-in muscular abdomen of the athlete. His face was twisted to one side by some early injury which had caused the necrosis of one side of his lower jaw. He was very kind to us children, and I remember his showing us a fool's cap with bells which he had worn at some lodge entertainment. He was scarcely thirty at that time, and single, though he later married and had a family.

My aunts had had better cultural opportunities than Uncle Jake. They had retained a good deal of Russian culture, though they later found that French culture had a greater commercial value for them in the garment trades. Aunt Charlotte, the mother of Olga, had a husband who had left her. She was divorced from him and was destined to know a similar misfortune again. Like her sister Augusta, who never married, she was in the needle trades. Both of them were exceptionally intelligent women; and with half an opportunity in life, they would have had careers comparable with that of my father. They spoke several languages fluently; and later on, when they had spent some years in Paris, they both became valuable assistants in the business of a New York couturier, where they passed themselves off as Frenchwomen.

Aunt Charlotte continued to work to a very advanced age and died only relatively recently from an accident. She was a very definitely Jewish-looking type, resembling some of those French Jewish women drawn with such zest by Du Maurier. Aunt Augusta rather resembled her, though she was much the better looking of the two. Like her sister, she also lived to an advanced age. There was an Aunt Adele as well, who later married and was a neighbor of ours for a brief time in the country, and still later moved out to the Pacific coast; but of her I have very little recollection.

There had been another brother, Moritz, older than my father, who had disappeared from the family view for many years. The last place from which he had been heard was Colon, or Aspinwall as it was then called, and the time had been one of a notorious outbreak of yellow fever. My grandmother always spoke of him as if he were still alive and might turn up at any time; but in her heart she knew that he had been long dead. Still, although it is rather late in the day, when my daughters wish to indulge in daydreams of sudden fortune, there is always the remote possibility that a very old gentleman may come to our house from a distant place, say in Australia, where he had made his fortune, and leave it to us in a burst of family sentiment.

As long as my grandmother lived, a visit to New York meant an unending series of courtesy visits to third and fourth cousins and their friends. I now know that this is an established part of the Jewish family structure, but at the time I did not even know that our family was Jewish. Of course, it was necessary for my mother to have some phrase in which to describe those qualities of my father's relatives which she did not want us to imitate; but the words "New York," spoken with a properly contemptuous intonation, was quite adequate to the purpose.

A child, however, is mostly interested in childish things, and the one of the family who was chiefly interesting to me was Olga. She showed me a few of the city child's tricks, such as putting pins on the streetcar tracks to be flattened out by the passing cars, and I believe that we used to play cassino together. Uncle Jake would show me card tricks, and how to build little houses out of old packs of cards. There were miniature children's cards that Olga used to buy in nearby stationery shops; the packs were always incomplete and there wasn't much to be done with them except to build houses.

The puffy little steam trains that ran over the Third Avenue Elevated delighted me. We used to travel downtown on it to make purchases in the big shops. One of my unpleasant memories is that of chasing around after my mother on these shopping errands, although there was no other way to outfit me for the ocean voyage to come, and many of the purchases were toys for the amusement of my sister Constance and me. I remember a little sailboat which I tried to sail on the lakes of Central Park. Neither my father nor I had the knowledge or experience to enable us to manage it.

Other presents for the voyage were sets of scientific experiments for children, entitled "Fun with Electricity," "Fun with Magnetism," and "Fun with Soapbubbles." I wonder whether the present generation, under the stimulus of Charles Addams, has extended the scope of these sets to cover "Fun with Atomic Physics," "Fun with Toxicology," and "Fun with Psychoanalysis." Whether they have or not, the sets of my childhood were thoroughly delightful, and even today I can remember the details of the experiments they contained.

At length we crossed the harbor by ferry to Hoboken, to embark in a ship of the Holland-America Line. We traveled second class, which represented the cheapest travel decently possible for a family with children in those days, when third class still meant steerage. Even in my early westbound trips after the First World War, I can remember looking down from the second-class deck onto the varied assemblage of steerage passengers, often still in their native costume, who herded together in a discomfort suggestive of the days of the Middle Passage.

A ship is a delightful place for a young child. There were a number of other children on board with whom I could play; and seasickness is chiefly a disease of the adult. I got into the proper amount of mischief, being duly chased out of the ship's working alleyway. I delighted in looking down from the deck on the ever-changing marbling foam. The ship's notices in English, German, and Dutch excited my interest, and I already knew enough German—I do not know how I learned it—to make out the similarity between the stateroom notices in German and Dutch. I unquestionably made a nuisance of myself to the passengers in the steamer chairs, as children always do on board ship, but I got properly punished for my sins when one passenger held me under his chair and tickled me past the point of all bearing. Finally, one morning I awoke to find the ship's engines stopped and a view of Rotterdam facing me through the porthole.

We took a compartmented European train to Cologne. I can still picture the railway station, the hotel at which we stayed, and the cathedral. The penny-in-the-slot machines of Germany were much bigger and finer than anything I had known at home, and the burnt almonds they sold were a new treat. Finally we went out to a suburb of Cologne where a cousin of my father lived.

I have said that I already had some fragmentary ideas of German before I left for Europe, but I doubt whether they would have added up to enough to allow my father to say that I knew any German at all. My father was a perfectionist in

CAMBRIDGE TO CAMBRIDGE, VIA NEW YORK AND VIENNA

languages, as befitted a man to whom languages came easily, and who had penetrated very deeply into them. His desire for the utmost finish and correctness was not always easy on his students, and was even harder on his family. My mother probably had a better than average ability to speak foreign languages and a fairly good acquaintance with German. Nevertheless, she was tongue-tied before Father's overwhelming proficiency. She admired his skill in languages, and let herself become unnecessarily dependent on it. As for me, it was not until I had left my home and married, and had come under the milder guidance of my wife, that I ever ventured to use a foreign language without a sense of guilt that led me to hesitate and to stammer each word.

To visit Europe with my father was to see it with the eyes of the European. Strictly speaking, I never went through that period of the tourist in which every door and every wall seems a fortress against him. For during this first visit my inadequacies as a stranger in the land were completely dwarfed by the different and greater inadequacies of the child. By the time of my second visit to Europe, when I was a young man, the memory of my first visit, my studies and reading, together with the continued presence of my father, had made Europe nearly as familiar to me as the United States. At no time could I thus contrast an unknown Europe with an America I knew well.

I will not say that I never went through any of the stages of brashness and bluster which belong to the innocent abroad. But the disease was brief, and it was greatly alleviated by my earlier inoculation with Europe. It has always seemed to me that Henry Adams, in his late attack of tourist frustration, was like a man who is first exposed to the mumps in his twenties. Adams remained allergic to modern Europe all his life. As for me, my early visit was perhaps the very best of all possible trainings, for the scientist must be a citizen of the world.

From Cologne we went up the Rhine on a steamer. We left the boat at Mainz, and made our way to Vienna. Vienna was our headquarters for a considerable period, and it is the part on our trip that has left the greatest impression on me. It is the little things that impress a child, and of these it is perhaps the smells which stay longest in one's memory. The smell of the alcohol lamp over which my parents prepared my sister's warm evening meals, the smell of the rich European chocolate with whipped cream, the smell of the hotel and the restaurant and *café*—all these are still sharp in my nostrils. I can remember the vegetarian restaurants at which we ate, which were generally up a flight or two in some obscure part of the city, and the skin on the boiled milk which I could scarcely bear to swallow. In Frankfurt we had tried a glass of *Apfelmost,* which had even worse consequences for me.

It was new for me to see the newspapers mounted on their wooden frames in the cafés. While Father would read his for the news, I would look over an English paper which contained a children's story as a *feuilleton.* My reading was not yet very fluent, and it was not easy for me to reconstruct the story out of the straggling portions which appeared from day to day and on which I could lay my hands, but I have a dim impression that the story was Kipling's "Puck of Pook's Hill." Certainly the dates are about right, since Kipling's story was written for his children when they were a trifle older than I was then, and they must have been born during his stay in Brattleboro slightly before my own birth.

One of my father's purposes in going to Vienna was to see a journalist by the name of Karl Kraus. I do not know what they discussed, though they probably concerned themselves with Jewish matters and quite possibly with the problem of the translation into literary German of the Yiddish poems of

CAMBRIDGE TO CAMBRIDGE, VIA NEW YORK AND VIENNA

Moritz Rosenfeld, the New York garment worker-poet, whom father had "discovered." I remember being taken into Kraus's apartment in an old-fashioned Vienna apartment house, and there I remember what seemed to me a confusion and disorder which I have never seen equaled elsewhere.

Vienna was hot and uncomfortable, and the bedbugs bit us children unmercifully. My parents did not know what had affected us, nor was the prominent dermatologist whom we called in of much use in the matter. He diagnosed our affliction as the itch, whose appellation of the "seven-year itch" did nothing to quiet the alarms of my parents.

When the true nature of the disease became manifest, we got no sympathy from our landlady. She informed us that in Vienna, that old city of stucco and crumbling plaster, no one was immune from them—not even the Emperor in his palace. They may have been old to the Emperor, but they were new to us; and they served a fully adequate notice on us to leave the city for a more salubrious place.

We found lodgings with a cobbler in the little Wienerwald town of Kaltenleutgeben. The house opened directly on the village street with scarcely any pavement in between. Behind it, the hill rose abruptly, and a little flight of steps led to a pleasant garden arbor. As in the case of the farmhouse where we had stayed in Alexandria, my early experience in Kaltenleutgeben was enough to fix the geography of the place sufficiently in mind so that I was able to recognize the house when I visited it more than thirty years later.

There were several boys in the cobbler's family, and they were my playmates. How we communicated with one another is a difficult question to settle now, for they certainly spoke no English, and my parents have assured me that I spoke no German—at least none by the standards of my father. That we did come to an understanding is clear, for we participated in more pranks together than did Max and Moritz, the young rascals whose history has been given us by Wilhelm Busch. When we were not examining the fat slugs and snails which peopled the back garden, we were playing forbidden games with the balls of the bowling alley attached to a neighboring restaurant, or decidedly not benefiting the cobbler's machinery by the liberties we took with it. We were to be found in the not too clean environs of the nearby open-air bath or buying little imitation baby bottles of colored water at the local fair.

Eventually, after a slow trip across Germany and Holland, we reached London. We found a vegetarian hotel in Maida Vale, which consisted of two large houses thrown into one, and behind it was a garden reserved for the abutting householders. I played with some of the children there, who, if I remember right, were the younger brothers and sisters of the famous pianist Mark Hambourg. The Boer War was still being fought at the time; and as my father was politically liberal, I echoed his opinions to my playmates by calling myself a pro-Boer, though I did not have the slightest idea what the conflict was all about. The English children retaliated by piling three-deep on top of me.

Not far from Maida Vale was the house of Israel Zangwill, with whom my father had corresponded about Zionist matters. Zangwill was one of the most eloquent British Zionists. My father foresaw the difficulties which have arisen since then from the superimposition of a Jewish colony upon a Moslem background. He was an assimilationist in a quite genuine sense, for he felt that the future of the Jews in the newer countries lay in their identifying their interests with those of the country, not in opening the wound of a separate new nationalism.

We visited Zangwill at his house near Maida Vale, which had a pretty little garden in front of it. He carried me upstairs on his shoulders. I remember his face: very Jewish, strongly lined, and not handsome but interesting and sensitive. I was to see him again on my next visit to Europe, which did not take place until I was eighteen years old.

He was not the only literary figure whom we visited in England. There was also Kropotkin, the great geographer and a genuine Russian prince of the imperial blood, who had turned anarchist as a young man, had tried to assassinate his cousin the Tsar, and had been obliged to flee the country. He had visited Boston about a year before and had been shown around by my father. He had given me as a present a little cardboard cabinet of minerals. One evening after he had been dined and wined by Mrs. Jack Gardner at her Fenway Palace, he turned up at our house, angry and inarticulate. "Wiener," he said to my father, "I have been insulted!" When Father had reduced him to comparative calm, the story came out. A Boston society lady had asked him, "Oh, Prince Kropotkin, how is your *dear* cousin the Tsar?"

We visited Kropotkin in his little house at Bromley, Kent. It was a workingman's home, with the usual depressing similarity to every house up and down the street. The back garden was a pleasant place, however, where his two daughters served us tea.

We saw the usual sights of London, such as the Houses of Parliament and Westminster Abbey. Sometimes we ate in a vegetarian restaurant on Holborn, and sometimes at an A.B.C. tearoom. We traveled mostly on busses, the upper deck delighting me; sometimes we took the new tube railway which in those remote days was known as the "tuppenny tube." The hansom cab still had no real competitor, and the London streets had what I later learned to recognize as the full, rich flavor of the Sherlock Holmes stories.

We left for Liverpool on the last lap of our journey, and of course the return voyage seemed as dull and uneventful as all such voyages, without the prospect of new adventure abroad to enliven them.

« V »

IN THE SWEAT OF MY BROW

Cambridge, September, 1901–September, 1903

ON AVON STREET in Cambridge we took a moderately old but pleasant house, with an ell behind and a little below the level of the main house. It had ground- and cut-glass front doors, a library and living room in front, and a small but adequate study for my father. The upstairs rooms were large and sunny, and the little upper story of the ell housed our nursery. There was a fairly large back yard for my sister and me to play in.

Some two houses from us lived Professor Bôcher, who, we later learned, was a great mathematician. He was the son of a former French professor of modern languages at Harvard, and I believe he had a family of two children about my own age. On the Easter of 1903, I joined his children to look for the Easter eggs which had been planted for their benefit. A little beyond him, well back from the road, was the house of Professor Otto Folin, the distinguished physiological chemist. Of Swedish peasant origin, he was married to an old-stock Western American woman, one of my mother's closest friends

IN THE SWEAT OF MY BROW

from her Missouri days. I had the run of their house and used to read their books. Both my mother and Mrs. Folin are still alive, and are still close friends.

The geneticist Castle and the physiologist Walter Cannon were two other friends of my father and of them I asked childish questions about science. My father and I went to see Cannon in his laboratory at the Harvard Medical School of that day, which was then behind the Boston Public Library, in a building now used by Boston University. I was particularly interested in the pictures Dr. Cannon showed us of the Canadian backwoodsman, Alexis St. Martin, who had accidentally shot a hole in his stomach, and of the American Army doctor, Beaumont, who had used him as a guinea pig for the study of digestion. Cannon himself told us the fascinating story of this partnership.

I was also interested in Dr. Cannon's X-ray machine, which, if I remember correctly, was excited by some sort of electrostatic generator. Cannon was perhaps the very first man to use the new ray of Röntgen in the study of the softer tissues, such as the heart and the stomach, and thus to continue the early work which St. Martin's ghastly fistula had made possible. He was also a pioneer in the use of lead screens for the protection of the X-ray operator. It was because of this precaution that he seemed for many years to have gone scatheless from these dangerous beams, while the majority of his early colleagues had crumbled to pieces by bits, submitting to amputation after amputation. Yet while he lived well into his seventies, his early X-ray burns killed him in the end.

These men I saw only occasionally. A much commoner visitor to our house was Father's friend, the Assyriologist Muss-Arnoldt. Muss-Arnoldt was, I believe, an Austrian Jew, and he had almost exactly the face and expression of his own Assyrian winged bulls. He was black-bearded and rather burly, a great scholar, and a man with an irascible disposition.

He taught me occasionally when he was staying at our house, and my father was otherwise occupied; and he was a strict but unskillful disciplinarian. One day a few years later, after a Latin lesson which particularly rankled in me, I was watering the lawn, and obeying a sudden and irrational impulse, I turned the hose on him. I was duly punished by my parents and Muss-Arnoldt looked askance at me ever after.

To a person who has seen the intervening stages of its development and decadence, it is difficult to compare the American Cambridge of today with the Cambridge of the beginning of the century. It is only by imperceptible steps that the houses have become grimier, that the traffic has become heavier, that the vacant lots have vanished, and that a community which in 1900 preserved much of the atmosphere of the country town has grown into a great, dirty, commercial city.

When I was a child, there were those who still spoke of Massachusetts Avenue by its old name, North Avenue; and it was lined by the inartistic but attractive and comfortable mansions of well-to-do businessmen. They are still standing, but fallen from glory. Their porte-cocheres shelter no coaches, and the elaborate wood carving of their porches is rotting away. They were inhabited by families with four or five children, and were ruled from the kitchen by a competent and masterful servant girl. The children had ample yards to play in, and the trees which shaded them had not yet been reduced to sickly pallor by the smoke of the East Cambridge factories.

The vacant lots of Cambridge bloomed with dandelions in the spring, buttercups in the summer, and the bluish blossom of chicory in the fall. The streets were, for the most part, unpaved; and when it rained, they were deeply rutted by the wheels of the horse-drawn delivery trucks. In the season of snow, the wagons were replaced by sleighs and sledges, and it was a favorite pastime of the youngsters to tie their sleds behind the delivery sleighs then known as pungs. On the hilly streets there was coasting, not only on the small sleds which one rode belly-bump but on large double-runners made up of two such sleds, a plank, and a steering wheel. There was an abundance of frozen puddles on which one could skate, and it was always possible to go to Jarvis Field and watch the Harvard hockey team at practice.

As I have said, my father was an enthusiastic amateur mycologist; and under his guidance, I toured vacant lots in search of morels in the spring and field agarics in the fall. The morels were confined to a few well-known spots, and the Harvard mycologists considered that they had duly staked their claims on these spots. It was a frequent cause of bad feeling when one of them stole a march on a colleague and reaped the little clump that the latter had considered his private property. Stands of field agarics were less subject to this test of ownership, and coprinus was too common to be considered a property at all.

These additions to our kitchen were supplemented by an occasional lepiota or a batch of elm mushroom. Every now and again we would find a clump of clavaria or hydnum, and even a few rarer delicacies; but these were mostly reserved for our summer vacations. Part of the fun was the fact that one might just possibly confuse these edible fungi with an amanita, or at least an emetic russula; and the knowledge that one would have to wait some twelve hours before the symptoms became obvious was a source of more than one sleepless night to my parents and to myself.

I have botanical memories beyond these stray fruits of the field. I can never forget the little maple keys taking root in the soil, nor the tiny trees which started from them. The smell of fresh earth, of maple bark, of the gum of cherry trees, and of newly mowed grass all belonged to my youth, with the drawl of the lawn mower and the pattering of water from the spray which kept our grass green. In the fall it was always delightful to trudge through the crisp heaps of fallen leaves in the gutter or to smell their aromatic smoke as they burned. In my childhood recollection, these are supplemented by the resinous perfume of freshly cut pinewood and the various builders' smells of linseed oil and new cement.

The whole frame of our lives has changed between that day and this. Wood then was so cheap that we used to knock up for firewood the boxes in which our groceries were delivered, and our butter came in wooden tubs or in neatly dovetailed wooden boxes with sliding lids. The chief token of those ampler days was, however, the ease with which one could secure servant girls. Mother never had less than two, a cook and a children's maid, together with the services of a laundress, and yet my father was only an impecunious instructor or assistant professor, with no promise of tenure for some time to come. For a large part of our time on Avon Street I nearly worshipped our maid, Hildreth Maloney, an intelligent, loyal and competent young woman, who was later to improve her position in the world. I do not remember our cook, but our laundress was a faithful and hard-working woman by the name of Maggy, to which we added the soubriquet, "The Buttonbreaker."

I was brought up in a house of learning. My father was the author of several books, and ever since I can remember, the sound of the typewriter and the smell of the paste pot have been familiar to me. But it was not the efforts of the literary scholar that first seized my imagination. By now I could read freely. I had full liberty to roam in what was the very catholic and miscellaneous library of my father. At one period or other the scientific interests of my father had covered most of the imaginable subjects of study. Somewhere in our bookcases there was a Chinese dictionary, there were grammars of unusual and exotic languages, there were charlatanlike books on the occult, there were accounts of the excavations of Troy and Tiryns, and there were a series of the English scientific primers of late Victorian times. Above all, there was a compilation of papers on psychiatry, electrical experiments, and travels of naturalists in the wilder parts of the world, that went by the name of the Humboldt Library. There were two odd volumes of the excellent *Natural History* of Kingsley, together with the far less scholarly and more anecdotic book of Wood which Mr. Hall had given me years before.

I was an omnivorous reader, and by the time I was eight I had overstrained a pair of rather inefficient eyes in consuming whatever books came my way. The learned works of my father's library shared my attention with the books of Dickens which my mother read to me, Stevenson's *Treasure Island*, *The Arabian Nights*, and the writings of Mayne Reid. To me they were all books of high adventure; yet the tale of Long John Silver and the stories in the *St. Nicholas Magazine* were pale to me beside the true accounts of the adventures of those naturalists who had found new beasts and birds and plants in the somber darkness of the rain forest, and had heard the raucous calls of the macaws and the parakeets.

Thus I longed to be a naturalist as other boys long to be policemen and locomotive engineers. I was only dimly aware of the way in which the age of the great naturalists and explorers was running out, leaving mere tasks of gleaning to the next generation. Yet even if I had been fully aware of this, my allegiance in science was already mixed. My father had brought me from the Harvard library a book devoted to the various branches of the study of light and electricity, which included a stillborn theory of television, frustrated by the inadequacies of the selenium cell. This book had attracted my fancy. I followed it up by further reading in physics and chemistry. When I was about seven years old, Father recognized this interest by inviting a chemical student, who had shown an interest in Russian and had attended his classes, to set up a little laboratory in the nursery and to show me some simple experiments.

Of course, I was particularly interested in the smellier of the experiments, and learned the trick of making a sulphide by heating scraps of metal with sulphur, and then of generating hydrogen sulphide by exposing this sulphide to the action of an acid such as vinegar. Mr. Wyman, my instructor, continued to teach me over the period of a few months later when I was forbidden to read because of my rapidly advancing myopia. Not long after this, I heard of his early death in an automobile accident not far from where my M.I.T. office is today. I believe that this was one of the earliest automobile fatalities in Cambridge.

Even in zoology and botany, it was the diagrams of complicated structure and the problems of growth and organization which excited my interest fully as much as the tales of adventure and discovery. Once I had been sensitized to an interest in the scientific—and various toys of scientific content played almost as great a role in this as my reading—I became aware of stimulating material all about me. I used to haunt the Agassiz Museum until there was more than one exhibit which I knew almost by heart. I read one scientific article that has had a direct influence on my present work, but I am unable to recollect where I saw it. It is confused in my memory with an article by Dan Beard which appeared in the *St. Nicholas Magazine* and was called "The Jointed Stick." It contained some material on the analogies and homologies of the skeleton of the vertebrates. The deeper article which my memory has long confused with this must have been written by some professional physiologist. It contained a very sound account of the progress of a nerve impulse along a nerve fiber, as a consecutive process of breakdown, analogous to the consecutive fall of a train of blocks rather than to a continuous

IN THE SWEAT OF MY BROW

electrical phenomenon. I remember that the article excited in me the desire to devise quasi-living automata, and that the notions I acquired from it survived in my mind for many years until they were supplemented in my adult life by a more formal study of modern neurophysiology.

Behind these books, which I read freely, there were a number that caused me a very real pain, yet a pain in whose titillations I was ashamed to observe elements of pleasure. No one had forbidden them to me, but I had forbidden them to myself, and yet when I turned past the fearful pages I could not refrain from giving them a stray glance. Much of *Struwwelpeter* came under this heading and a good deal of *Max und Moritz*. In *The Arabian Nights* there is a terrible "Tale of the Greek Physician," and there is Grimm's fairy tale of "The Boy Who Did Not Know Fear." There were certain parts of the scientific books to which I had access that excited this baser mixture of emotions, and I remember in particular terrifying but fascinating passages in the Humboldt Library which were devoted to an account of execution by electricity and of fashion in deformity. I had an early interest in medical books which was partly legitimate and scientific but which also contained not a little element of "looking bogy in the face." I was quite aware of the mixture of emotions with which I read these, and I was not able for any length of time to pretend that my interest was altogether innocent. These books aroused or recalled emotions of pain and horror, yet showed these emotions to be related to those of pleasure. I knew this then, long before Freud's work had come to my attention and helped me to understand these tangled emotions.

Probably much of my early reading was over my head at the time. It is not essential for the value of education that every idea be understood at the time of its accession. Any person with a genuine intellectual interest and a wealth of intellectual content acquires much that he only gradually comes to understand fully in the light of its correlation with other related ideas. The person who must have the explicit connection of his ideas fed to him by his teacher is lacking in the most vital characteristic that belongs to the scholar. Scholarship is a progressive process, and it is the art of so connecting and recombining individual items of learning by the forces of one's whole character and experience that nothing is left in isolation, and each idea becomes a commentary on many others.

This unusual reading history of mine made me difficult to place in school. At seven, my reading was far in advance of my handwriting, which was awkward and ugly. My arithmetic was adequate but unorthodox, in that I preferred to use such shortcuts as to add nine by adding ten and subtracting one. I still was inclined to do sums on my fingers, and was not yet very sure of the later parts of my multiplication tables. I had the beginnings of a familiarity with German, and I devoured every scientific book on which I could lay my hands.

After a certain amount of looking around, it was decided to put me in the third grade of the Peabody School on Avon Street. The teacher was kind and intelligent, as well as very tolerant of my infantile maladroitness. I do not know how long it was before my parents and my teachers came to the conclusion that I should be shifted to the fourth grade. I do not believe that they waited all year to make this decision. I still could scarcely have been much more than seven years old at the time. At any rate, the fourth grade teacher was less sympathetic with my shortcomings, and in one way or another I did not click.

My chief deficiency was in arithmetic. Here my understanding was far beyond my manipulation, which was definitely poor. My father saw quite correctly that one of my chief difficulties was that manipulative drill bored me. He decided to take me out of school and to put me on algebra instead of arithmetic, with the purpose of offering a greater challenge and stimulus to my imagination. From this time until I went to the Ayer High School at the age of nearly ten and even later, all my teaching was in my father's hands, whether directly or indirectly.

I do not think that his original purpose had been to push me. However, he had himself started his intellectual career very young, and I think that he was a little surprised by his own success with me. What had started as a makeshift was thus continued into a definite plan of education. In this plan, mathematics and languages (especially Latin and German) were central.

Algebra was never hard for me, although my father's way of teaching it was scarcely conducive to peace of mind. Every mistake had to be corrected as it was made. He would begin the discussion in an easy, conversational tone. This lasted exactly until I made the first mathematical mistake. Then the gentle and loving father was replaced by the avenger of the blood. The first warning he gave me of my unconscious delinquency was a very sharp and aspirated "What!" and if I did not follow this by coming to heel at once, he would admonish me, "Now do this again!" By this time I was weeping and terrified. Almost inevitably I persisted in sin, or what was worse, corrected an admissible statement into a blunder. Then the last shreds of my father's temper were torn, and he addressed me in a phraseology which seemed to me even more violent than it was because I was not aware that it was a free translation from the German. *Rindvieh* is not exactly a complimentary word, but it is certainly less severe than "brute"; and *Esel* has been used by so many generations of German schoolteachers that it has almost become a term of endearment. This cannot be said of the English word "Ass!" or of its equivalents, "Fool! Donkey!"

I became accustomed to these scoldings quite rapidly; and

in view of the fact that my lessons never lasted many hours, they were emotional hurdles which I could take in my stride. However, they never ceased to be genuine hurdles. The schoolmaster everywhere can summon to his aid the absurdity of his pupil. The very tone of my father's voice was calculated to bring me to a high pitch of emotion, and when this was combined with irony and sarcasm, it became a knout with many lashes. My lessons often ended in a family scene. Father was raging, I was weeping, and my mother did her best to defend me, although hers was a losing battle. She suggested at times that the noise was disturbing the neighbors and that they had come to the door to complain, and this may have put a measure of restraint on my father without comforting me in the least. There were times for many years when I was afraid that the unity of the family might not be able to stand these stresses, and it is just in this unity that all of a child's security lies.

But much more serious for me were the secondary consequences of my father's discipline. I used to hear my juvenile ineptitudes repeated at the dinner table and before company until I was morally raw all over. On top of this, I was made well aware of the shortcomings of my father's father, and it was borne in upon me that his worst traits were latent in my makeup, and only waiting for a few years to be brought out.

When I now read John Stuart Mill's account of his father, it seems on the surface to have represented a completely virtuous relationship on both sides. I know better, and when I read his few words about his father's irascibility I know just how to interpret these statements. I am certain that even if that irascibility had been more decorous than that of my father, it had probably been no less unremitting. There is passage after passage in Mill which could well be the statement by a proper Victorian of a course of training which had been very close to that I had experienced.

IN THE SWEAT OF MY BROW

My own education had both remarkable similarities to that of Mill and important differences from it. Mill's education was predominatingly classical at a time when there was no other basis for a sound training. Hence Mill covered a wider range of the classics than I did, and at an earlier age; but he began mathematics rather later, and his father was a less authoritative preceptor in these matters. My father had shown from his youth a rather outstanding mathematical ability, which he imparted to me from my seventh year on. Moreover, by the time I was seven, my own reading had penetrated into branches of biology and physics which were even beyond my father's own scope, and which must have gone far beyond the rather pedantically classificatory natural history available to the boy Mill during his tramping excursions.

In one respect my father resembled James Mill: both were ardent walkers and loved the countryside. I gather, however, that the elder Mill did not have the green thumb of which my father was so proud, and that the boy was not under the same pressure to work in the garden and in the field. With Mill as with myself, walks with our fathers seemed to be a fruitful source not only of outdoor pleasure but of the moral stimulus derived from contact with men of learning and of character.

Both the Mills seem to have centered their lives around questions of ethics. They were of a Scottish family, and it is every Scotsman's birthright to be a philosopher and a moralist. It is likewise the birthright of every Jew. And yet the more impulsive character of the Mediterranean gives to his philosophizings and moralizings a different appearance from that which belongs to the man of the north.

The Mills rank as two of the great humanitarians of history. My father's career shows an almost equal depth of humanitarian motivation. Yet the roots of his humanitarianism were different from the Mills', as different as Jeremy Bentham and Leo Tolstoy. The Mills' passion for mankind was an intellectual passion, full of nobility and righteousness, but perhaps rather arid in its lack of an emotional participation with the oppressed. The roots of my father were in the deep human sympathy of Tolstoy, which itself has much of the compassion and self-abnegation of the Hindu Holy Man. In short, the Mills were classicists partaking of the sympathies of a romantic period, whereas my father, although educated in the classical tradition, was a romanticist of the romanticists.

I cannot imagine my father or myself being greatly moved as the Mills were by the icy glitter of Pope's translation of Homer. The poetry that most moved my father, as it has most moved me, was that of Heine, with its aspiration for the beautiful and the bitter revulsion which comes as the poet sees far too clearly the horrible contrast of that which is with that which he would like to believe. I cannot imagine Mill regarding Heine as more than an impertinent upstart, although there well may be hidden references to Heine in Mill's books which give me the lie.

In the details of Mill's experience and of my own as well as in their larger lines, there is much that is parallel. It is clear that both of our teachers wished to prevent us from taking ourselves too seriously by a policy of enforced modesty, which at times amounted to systematic belittling. It is clear that both children combined a profound respect for their fathers with a certain degree of an inner feeling of deprivation and resentment. Yet the conflict of son and father has come to show itself in very different ways. There seems to have been in both Mills an aversion to any display of emotion, which was certainly not present in my father. Yet it is quite clear from Mill's account of his training that strong emotions were there and that they were not weakened in any way by the impassive façade which both father and son maintained.

I doubt whether the older Mill had any of the possibilities of explosiveness and anger that certainly existed in my father, and I equally doubt that he showed the human weaknesses and longings which at times almost reversed the roles of father and son in my family, and made me love my father the more deeply because he never wholly ceased to be a child. In Mill's book it always seems that the awareness of his ambivalence towards his own education is pruned like the trees of an eighteenth-century garden.

That we may readily be aware of the suppressed conflict between John Stuart Mill and his father we owe in part to Samuel Butler. Samuel Butler was perhaps not a prodigy in the full sense of the word, but like many of the infant prodigies, he had been brought up under the intimate supervision of a dominant father, and like many of the infant prodigies including myself, this supervision had led to a certain degree of revolt in his reminiscent attitudes. Indeed, I feel that Samuel Butler as the Ernest Pontifex of *The Way of All Flesh* suffered from a parental tutelage at least as strict as my own, and at the hands of a man infinitely more commonplace and less sympathetic than my father. His ambivalence toward his own father had much more in it of hatred than of love, and what respect it contained was more a respect for strength of character than for good will. I cannot deny that in my own attitude to my father there were hostile elements. There were elements of self-defense and even fear. But I always recognized his exceeding ability in intellectual matters and his fundamental honesty and respect for the truth, and these made tolerable the many frequently occurring painful situations which must have been absolutely intolerable to the son of a Rev. Mr. Pontifex.

As far as the impact of the outer world is concerned, the conventionality of the parental Pontifex certainly offered Ernest a most intense situation of conflict, but it spared him from something of the potential disapproval of the world about him as a breakwater spares the ships of the harbor. The Rev. Mr. Pontifex was unconventional in nothing but the massiveness and whole-souledness of his conventionality. For myself, with all the understanding of my father, I had to pay the double penalty of being the unconventional child of an unconventional man. Thus I was isolated from my environment by two separate isolations.

Religious problems seem to have dominated Samuel Butler and also John Stuart Mill in the relations with their fathers. These problems were even more acute in the youth of Edmund Gosse, another writer who must be mentioned in the discussion of father-son relations. Gosse's book *Father and Son* is like Butler's in being the account of the relations of a boy with a desire for independence to a very dominant father with theological interests. Indeed, Mill's book, for all of the want of a formal theology on the part of both father and son, has a strongly ethical tone which echoes similar preoccupation. In my own case, while my father was a man of strong moral sense, it cannot be said that he had any great interest in theology. The source of his humanitarianism was Tolstoy, and even though Tolstoy embellishes his propagandist texts with many quotations from the Bible, he is at home with that side of Christianity which preaches humility and charity and extols the virtue of the oppressed and undervalued. I have already said that I had begun to express doubts of religion at the early age of five, in terms that would have brought me severe castigation and even more severe chiding at the hands of the elder Butler or the elder Gosse.

Let me return to the details of my own history. I certainly do not remember any effective opposition on the part of my father. Indeed, I strongly suspect that my infantile adventures in agnosticism and atheism were scarcely more than a reflection of my father's own attitude which may have reflected the attitude of my scapegrace grandfather, who had already left the fold of Judaism without embracing any equivalent religion. Even a skeptic like James Mill would have found my levity intolerable. My own career as an infant prodigy thus differs from that of these victims or beneficiaries of dominant fathers in that it was entirely on a secular plane.

It is clear that religion or the equivalent moral questions were what made the mid-Victorian tick. With my father as with me, the predominating motive was that of a profound intellectual curiosity. He was a philologist; and for him, philology was more nearly an exploratory tool for the historian than a declaration of learning, or the means of taking to one's own soul the great writers of the past. Although there was always a strong moral implication in my father's personality and in the course of life toward which he directed me, my interest in science started with a devotion rather to the service of truth than to the service of humanity. Such interests in the humanitarian duties of the scientist as I now have are due more to the direct impact of the moral problems besetting the research man of the present day than to any original conviction that the scientist is primarily a philanthropist.

The service of truth, though not primarily a task of ethics, is one which both my father and I conceived to impose upon us the greatest moral obligation possible. In a later interview which my father gave to H. A. Bruce, he stated this in his own words.[1] The legend that Galileo after his conviction was heard to say, *"Eppur'si muove!"* ("But it *does* move!"), while apocryphal, is true in essence as depicting the code of the scientist. My father felt the demand of intellectual honesty to be one which the scholar can as little repudiate on the basis of any personal danger into which it might lead him as the soldier can repudiate the duty to fight at the front or the doctor to stay and be effective in a plague-stricken city. Never-

[1] *The American Magazine*, July, 1911.

theless, it was an obligation which both of us conceived to belong to a man, not merely as a human being, but precisely because he had chosen himself for the specific devotion of being a servant of the truth.

I have said that my father was a romanticist rather than a Victorian classicist. His closest spiritual kin, besides Tolstoy and Dostoevsky, were the German Liberals of 1848. His righteousness partook of the element of *élan*, of triumph, of glorious and effective effort, of drinking deep of life and the emotions thereof. For me, a boy just starting life, this made him in many ways a noble and uplifting figure, a poet at heart, amid the frigid and repressed figures of an uninspiring and decadent Boston. It was because of this, because my taskmaster was at the same time my hero, that I was not bent down into mere sullen ineffectiveness by the arduous course of discipline through which I went.

My father not only taught me directly but also had a Radcliffe pupil of his, Miss Helen Robertson, come several times a week to review my Latin with me by ear and to help with my German. It was a delight to have her come and to have opened to me another contact with the world of grownups besides that of my family. I learned from her the legends of Harvard and of Radcliffe; of the acerbity of this professor, of the wit of that; of the old peddler known as John the Orangeman and the cart which the Harvard students had given him with the donkey bearing the name of Ann Radcliffe; and of the wonderful blind-and-deaf student, Helen Keller. I learned of the visit of Prince Henry of Prussia and of the student pranks on that occasion. In short, even at the age of eight, I had a foretaste of the life of a college student.

It was about this time that I began to discover that I was clumsier than the run of children about me. Some of this clumsiness was genuinely poor muscular co-ordination, but more of it was based on my defective eyesight. I thought that

IN THE SWEAT OF MY BROW

I could not catch a ball, when the really fundamental fact was that I could not see it. Undoubtedly all this was accentuated by the early age at which I learned to read and by my immoderate indulgence in that pastime.

My appearance of clumsiness was accentuated by the learned vocabulary which I had acquired from my reading. While it was entirely natural and not in any degree an affectation, it emphasized to my elders and particularly to those who did not know me very well that I was in some sense a misfit. As I shall point out in the next chapter, I had a fairly normal acquaintance with other boys of my age, so that I do not believe this anomaly in fact excited as much attention among my contemporaries as it did among my seniors. If my contemporaries received any particular impression from my adult vocabulary, I am inclined to believe it was only a secondary impression conveyed to them by their parents.

During the year when I was eight, my eyes began to trouble me in a rather alarming way. Of course my parents noticed this long before I did. A child is not aware of a constant deficiency of sense such as of eyesight. He accepts his own vision as the norm of vision, and if there are any defects, he assumes that they are common to the human race. Thus while a rapid aggravation of eye trouble is noticeable, a steady level of visual deficiency calls no attention to itself, especially when, as is the case with the myope, the difficulty does not interfere with reading. The myope tends to hold the book too close to his eyes, and this is conspicuous to his more sophisticated parents. But it is not conspicuous to himself until it has been pointed out, and until he has been given the advantage of adequate glasses.

My parents took me to Dr. Haskell, our oculist, who gave strict orders that I was not to read for a period of six months, and that at the end of this period the entire question of my reading was to be reconsidered. Father went ahead teaching me mathematics, both algebra and geometry, by ear, and my chemistry lessons went on. This period of ear training rather than eye training was probably one of the most valuable disciplines through which I have ever gone, for it forced me to be able to do my mathematics in my head and to think of languages as they are spoken rather than as mere exercises in writing. Many years later my training proved of great service to me when I came to learn Chinese, which a complicated notation has rendered far more difficult to the eye than it is to the ear. I don't suppose that this early training created the very good memory which I have carried with me down to the present day, but it certainly showed me that I had such a memory and made it possible for me to exploit it.

At the end of six months, my myopic eyes showed no further alarming symptoms, and I was allowed to read once more. The doctor's judgment in permitting me to go back to my work has been justified by the last fifty years of my life, for in spite of increasing nearsightedness, cataracts, and the removal of both lenses, I still have very fair vision, and I see no prospect that my eyes will let me down as long as I live.

There is one particular passage in Mill's *Autobiography* which excites a certain resonance in my own experience. Mill speaks of passing on his instruction to his younger brothers and sisters. My sister Constance tells me that she suffered much from my juvenile didacticism. I certainly was not made the official pupil-teacher in my family as Mill was. Yet the entire example of a life in which the person one most respects always appears as a teacher can only make the child think of maturity and responsibility as the maturity and responsibility of the school master. It is inevitable that all concentrated teaching teaches the boy to be a teacher. This may be overcome later, but it represents a trend that must always be there.

During the next years, without excessive difficulty but with a severely lacerated self-esteem, I labored under my father's

IN THE SWEAT OF MY BROW

tutelage through the Wentworth textbooks on algebra, plane geometry, trigonometry, and analytical geometry, and learned the rudiments of Latin and German. I recognized that my father spoke with the authority of the scholar, even as I recognized that most of my outside teachers had spoken with something less.

« VI »

DIVERSIONS OF A WUNDERKIND

THE LAST CHAPTER was devoted to my work in the early days of my career as a *Wunderkind*. However, my life was one of play as well as work. My parents entered me as a member of a playground which had been set up in a vacant lot next to the Peabody School. We had to show a card to get in and to allow us to use the services of the playground teacher as well as to crawl through the jungle gym and to coast down the slide or to employ such other devices as were there for our use and our exercise. I spent much time there talking with the policeman on the beat. Patrolman Murray lived opposite us and he loved to tease me with tall tales of police service.

I had many playmates whom I found at the Peabody School and retained even after my father had taken over my education. There was Ray Rockwood, who later went to West Point and died many years ago as an officer in the service. He was endowed with two aunts, whose efforts mutually canceled each other. One was a Christian Scientist, and the other manufactured some sort of proprietary medicine.

DIVERSIONS OF A *Wunderkind*

Walter Munroe was a son of a starter of the Boston Elevated Railway, and Winn Willard was a carpenter's son. Another of my playmates was the son of a man who later became mayor of Cambridge. The King boys, the sons of a Harvard instructor, were mechanically gifted and owned a little working steam engine which was my envy. *The Youth's Companion*, for which my parents then subscribed on my behalf, offered such engines among the premiums in their subscription contests, but even without competing, it was possible to buy the premiums through their services at a reduced rate. My parents bought many toys for me in that way, but they never went quite so far as the steam engine.

In those days the papers were full of what was then an unfailing source of news; the persecution of the Armenians by the Turks. How we came to the conclusion that it was any of our business, I do not know, for we certainly knew little about Turkey and less about Armenians. One day the King boys and I decided to run away to the wars and fight on behalf of the oppressed. How my father got on our trail I do not know, but in about a half hour he found three very confused little boys gazing into a shop window on Massachusetts Avenue half way between Harvard Square and Central Square. He delivered the King boys to the mercy of their own family. To me he administered no punishment except that of a biting ridicule. It was years before my parents stopped teasing me about this occurrence, and even to the present day the memory of this teasing can hurt.

Most of the survivors among my childhood playmates have made good in the world. One of them, who was notorious among us all as a particularly nasty and vicious child, is now a great tycoon of industry. Another, who distinguished himself by chasing a comrade through the streets with a hatchet, has disappointed all of us by eschewing the life of violence for the scarcely more satisfactory career of a petty swindler.

EX-PRODIGY

We had all sorts of fights in those days, from snowball fights to a serious gang affair in which two armies of boys met on Avon Hill Street and pelted one another with stones. Our parents soon broke this up. In one snowball fight a companion of mine, who suffered from a high degree of nearsightedness, incurred a detached retina and lost the sight of one eye.

I have said that I, too, was a myope, and I suppose it was this snowball accident as much as anything else that led my parents to punish me for fighting and otherwise discouraged fighting at all costs. I never would have made a good fighter, as the effect of any severe emotion was to paralyze me with such weakness of fear that I could scarcely utter a word, let alone strike a blow. I suppose the reason was as much physiological as psychological, as I have always gone into fits of weakness when my blood sugar was low.

I took a sufficient part in the sports of the children of my age. I helped to make snow forts for snowball battles, as well as the snow prisons in which we immured our captives and in which I occasionally got immured myself. I jumped on behind the delivery sleighs or "pungs" which traversed the yellow slush-covered streets of the winter Cambridge of those days. I scaled the back fences with the best of them, and ruined my clothes when I fell off. I tried to skate on a child's double-runner skates, but my ankles were weak and lax, and I never graduated to the more efficient single runners. I coasted down Avon Hill Street and would try to persuade my seniors and betters to give me a ride on their swifter double-runner sleds. In the spring I searched the pavements and the yards for little pebbles which I could grind up with spittle to make a crude sort of paint, and I would chalk the pavements to make hopscotch courts on which my comrades and I could play. I walked over to North Cambridge to get comic valentines or Christmas cards from the stationery shops, according to the

DIVERSIONS OF A *Wunderkind*

time of year, as well as cheap candies and the other delightful trifles of extreme youth.

I used to play a great deal with miniature electric motors. At one time I had the vision of making one of these, following the directions in a book which I had received as a Christmas present. However, the book was written from the point of view of the boy who has a small machine shop at his disposal; and even if I had possessed one, I neither then nor later would have had the mechanical skill to make use of it.

I remember among my toys a megaphone, a kaleidoscope, and a magic lantern, as well as a series of magnifying glasses and simple microscopes. The magic lantern had a number of comic slides with it, which were quite as gratifying to a small boy of that day as is a Walt Disney movie to his present successor. We used to hold magic lantern shows in the nursery and to take our pay in pins.

There were times when we tried to make a little real money for our undertaking. Father had a series of photographs of Greek art which I understood had been given to me, and I tried to sell them around the neighborhood. I had a pretty task to collect them when my parents found out what I had done.

Christmas of 1901 was hard for me. I was just seven. It was then that I first discovered that Santa Claus was a conventional invention of the grownups. At that time I was already reading scientific books of more than slight difficulty, and it seemed to my parents that a child who was doing this should have no difficulty in discarding what to them was obviously a sentimental fiction. What they did not realize was the fragmentariness of the child's world. The child does not wander far from home, and what may be only a few blocks away is to him an unknown territory in which every fancy is permissible. These fancies often become so strong that even when the child has penetrated beyond the previously unknown boundaries, the conviction of his imagination maintains him in accepting a geography that his experience has already shown him to be false.

What is true concerning the physical map is also true concerning the chart of his ideas. He has not yet had the opportunity to explore very far from the few central notions that are his by experience. In the intermediate regions, anything may be true; and what is for his elders at least an emotional contradiction is for him a blank which may be filled in any one of several ways. For the filling of much of this blank he must depend on the good faith of his parents. Thus the breaking of the Santa Claus myth discloses to him that this dependence on the good faith of his parents has its limitations. He may no longer accept what they have told him, but must measure it by his own imperfect criteria of judgment.

The family was enlarged again in the spring of that year. My sister Bertha was born, and her birth nearly cost my mother's life. Our neighbor, Dr. Taylor, attended my mother. He was a gray-bearded, elderly man, with two sons who were among my playmates. As before, Rose Duffy was the midwife. I was full of fancies about what birth might mean, and had a weird idea that if one could put a doll, say a doll made out of a medicine bottle, through the proper course of incantation, one could make a baby out of it.

This naïveté was remarkable in view of my scientific sophistication at the time. The various biological texts which I read between my sixth and my ninth years contained a great deal of material on the sexual phenomena of animals in general and of vertebrates in particular. I was quite aware of the main outlines of mitosis, of the reduction divisions of the egg and the spermatozoon, and of the fusion of male and female pronuclei. I had a fair idea of the elements of embryology and of the gastrulation of some of the lower invertebrates. I knew that these facts were somehow connected with human reproduction, but my inquiries of my parents in that direc-

DIVERSIONS OF A *Wunderkind*

tion were not encouraged, and I was quite aware that at some place in my line of thought there was a clue missing. Intellectually, I was far advanced in the understanding of the phenomena of sex both in plants and animals. But emotionally the whole matter was as indifferent to me as it can only be to a young child: or rather, where it was not a matter of indifference to me, the only emotions it excited were those of puzzlement and terror.

What made the family situation even worse at Bertha's birth was that both my sister Constance and I came down with measles about the time of mother's delivery. I don't remember how we managed to take care of the three of us at the same time.

It was about this time that my parents tried to see if I could be brought into greater conformity with the habits of the other faculty children. They sent me to a Unitarian Sunday school after a considerable amount of protest which I took out in philosophical debate with the minister after Sunday school. The minister was Dr. Samuel McCord Crothers, that admirable essayist and litterateur, who was a friend of the family for many years and who more than twenty years afterward officiated at the marriages of my sisters. Dr. Crothers was not shocked by my youthful rejection of religion, and tried to meet my arguments seriously. At any rate, through his forbearance it was not absolutely impossible for me to continue in Sunday school.

The Sunday school had a good library, and there were two books which I remember impressed me particularly. One was Ruskin's *King of the Golden River*. Many years afterward, when I read his *Modern Painters*, I recognized the same sense for mountain scenery and the same strong ethical attitude which I already knew in his story for children. The other book was an English version of a French story of the seventies entitled *The Adventures of a Young Naturalist in Mexico*. It is only within the current year that I have seen this book again and I have renewed my impression of the rich picture it gives of the lushness of the tropical forests of the Mexican lowlands.

The Sunday school gave a Christmas play in which I was due to appear somehow or other in a minor part. The making up and dressing up embarrassed me exceedingly, and created a disgust for participating in amateur dramatics which has lasted to this day.

That summer, which we spent in a cottage in Foxboro, *Cosmopolitan Magazine* published the serial story by H. G. Wells entitled, "The First Men in the Moon." My cousin Olga and I devoured it, and although I was not able to appreciate all the social significance of the writing, I was properly shocked and terrified by the brittle figure of the Grand Lunar. About the same time, I had been reading Jules Verne's *The Mysterious Island*. These were the two books which introduced me to science fiction. Indeed, for many years I remained an *aficionado* of Jules Verne, and a trip to the library to find yet another volume of his writing was probably a greater delight than this generation of children can get out of the movies.

Parenthetically, for all this I am not enthusiastic over modern science fiction. Science fiction has been rapidly formalized and it is no longer a genre which offers sufficient freedom for the author who tries to follow its accepted canons. I have tried a little fictional writing about scientific matters, but it is entirely outside the frame of the science fiction monopoly. Some writers in this field have let their taste for fiction outstrip their sense for fact, and have allowed themselves to be used as promoters for various schemes of charlatans. The very originality of science fiction has become a cliché. Its slickness is quite different from the enthusiasm and verve with which Jules Verne adapted the romantic milieu of

DIVERSIONS OF A *Wunderkind*

Dumas, or the sincerity by which H. G. Wells made his sociological discourses palatable and fascinating.

Whether it was summer or winter, Father did a considerable amount of literary work, and I always found it very exciting to follow the successive stages of publication. The first book of his own, as contrasted with the *Poems* of Moritz Rosenfeld (which he had helped to see through the press), was a *History of Yiddish Literature*. This was a little too early for me to have clear memories, but I well remember his next book, the two-volume *Anthology of Russian Literature*, which he edited and the component parts of which he in large measure translated. This was followed by a great contract with Dana Estes and Sons, by which my father agreed to translate all the works of Tolstoy for the cash sum of ten thousand dollars. This was a rather skimpy reward even at that time, and today it seems a ridiculously small sum to pay for the translation of twenty-four volumes. My father accomplished this task in twenty-four months. In this he was helped by a very competent secretary, Miss Harper, and I believe that she was paid directly by the publishers. Father's relations with his publishers were never very smooth, and I think that he was justified in his general attitude of suspicion.

I soon learned that a manuscript is followed by the long banners of galley proof, and these in turn by the smooth oblongs of page proof and the leaded oblongs of plate proof. I learned the main signs of the proofreader and the general technique of proof correction. I learned that authors' corrections in galley proof are expensive, while they are exorbitant in page proof and practically prohibitive in plate proof. I saw Father cut up two or three Bibles in order to translate the Biblical quotations of Tolstoy, and I used to play with the discarded proof sheets and remains of these Bibles as if I were reading proof myself.

Although I had met my mother's family before we moved

to Avon Street, most of my recollections of them belong to that time. My mother's mother and two of her sisters followed her to Boston at some period which I do not remember. My grandmother lived in a Cambridge lodging house on Shephard Street at the time that my sister Bertha was born; and I can still remember a heroically determined effort on her part to bathe me, in which she showed herself quite indifferent to my imminent suffocation and to the action of soap on my eyes.

I do not think that she had any particular quarrel with me, but she certainly did with my parents. I do not know in what manner the latter had offended, although it seems clear that the old quarrel between German Jew and Russian Jew played at least a role. At any rate, my parents accused my mother's family of trying to break up their marriage, and there followed one of those family feuds that are not even terminated by death. Some of the participants in these feuds may die, but the rancor of the survivor lives with their memory.

I met my Grandfather Kahn only once if at all. I know his looks very well from his photograph, which is that of a tall, grave man with a long, grey beard. He was already separated from my grandmother, and lived in some sort of old people's home in Baltimore. I remember that at some later birthday he sent me a gold watch for a present. He died about 1915.

My father and I spent a great deal of time in the spring of 1903 looking for a place to pass the summer. We made a regular circuit of the villages south of Boston, from Dedham to Framingham, and even along the seacoast around Cohasset, but we never found the right place. We asked the advice of all of father's friends who lived in the outer suburbs. Finally, we decided to look a little farther afield in the northwest sector, and hit on a place called Old Mill Farm in the town of Harvard, about halfway between Harvard Village and Ayer Junction. We spent one summer there getting acquainted with the place, and decided that the next summer should see us engaged in modernizing the farmhouse and preparing to lead the simple life of a farmer and a college professor.

I do not know precisely what relation the name of the town of Harvard has to Harvard University, but any connection between the two places themselves is remote. The town of Harvard is distinguished historically for containing the site of the first or second water-driven grist mill built in inland Massachusetts. Though this mill was not on the farm we eventually bought, the old dam still stood near its boundaries, and the pond had been successively enlarged by later and later structures until the dam had come to stand opposite the farmhouse. Hence the place was called Old Mill Farm, and this is the name by which I shall refer to it in the ensuing chapters.

When Father bought Old Mill Farm and decided to live there later all the year round, I think there were several motives at work. One was his love of the country and his desire to work in the soil. Another (which I think must have been much less important) was the pride in the additional status of the landowner. Without any doubt, Father considered it essential for his children to have as much of their bringing up as possible in the country, and I believe he found my schooling problem less unsolvable than it might have proved in the city where the only choice would have been either a rather rigid public school or a rather expensive private school. I don't think Father could have found the country more conducive to his literary and scientific work than the city, and indeed it was pretty obvious to me that he made a considerable sacrifice in commuting as he did between Ayer and Cambridge.

When we first came to Old Mill Farm in the summer of 1903, the farmhouse was a gaunt, unattractive structure, dating from the decade before the Civil War. The house stood gable-end to the road, and was connected with a large barn by the usual sequence of ells and woodsheds. Opposite the house was the pond, which then seemed to me almost a lake but which could scarcely have been more than two hundred feet wide. It had a marshy island in it and a little grove of trees on the right side, in which we found ferns and trilliums in the early summer. On the other side was the dam, from which two streams led across a boggy meadow and under the road to the extreme limits of our farm. Beside one of the streams, and nestling under the dam, was a shed equipped with a turbine; it had been used by a previous owner as a small factory for some product I cannot remember.

The land between the two streams and the road was a tangle delightful to a youngster. There were frogs and turtles in the streams. The little fox terrier who was my personal pet soon learned that I was interested in them, and would retrieve turtles for me between his jaws. The tangle of weeds in the half-marshy triangle of land was rich in flowers interesting to the child, such as touch-me-not, joe-pye weed, turtle mouth, and spiraea. Down from the stone embankment which carried the road hung festoons of the vine of the wild grape. The meadows were full of blue, yellow, and white violets, of wild iris, and of bluets and sweet grass after their season. In a more remote pasture there grew the two gentians, fringed and closed, as well as both the pink and the white spiraea and an occasional bush of rhodora.

All these were delightful to me. Not less so were the willows which lined the pond, together with an old stump grown up with withies, which formed our playhouse. There was a nearby sandpile, where we pitched a tent made of old rugs and piano boxes. By the sandpile were needle-covered banks under a spreading pine tree, and there we could burrow and make little ovens in which we baked potatoes. The sandpile was the washed-out part of an old road which had led past our house before the present one had been located, and down which Lafayette is said to have ridden on his great tour about the United States, when he came back as the guest of the country. A trail from the sandpile led down through a wet wood of alders to the sandy shore of the lake, where my sister and I used to bathe among tadpoles, leeches, and tiny frogs, before we had learned to swim enough to be trusted off a shelving beach. Later on, when we were older, our favorite bathing place was a pool just above the great dam, where the main stream poured over in a waterfall, and I could just stand up on tiptoe with my nose out of the water.

There was a boat on the pond, and we used to row up past the ruins of the seventeenth-century dam well into the inlet of the pond. With its water lilies, yellow and white, its pickerel weed, its bladderwort, and the mysteries of its turtles, its fish, and its other submerged inhabitants, the pond was always a delightful place for us. So was the old henhouse, with its chicken-wire supported on live willow posts which had struck root and had grown into young trees. So was the barn with its hay loft, where one could hide and slide and jump to one's heart's content. So were the neighboring farmhouses, at whose back doors we always stopped for a glass of cold water and a pleasant word with the farmer's wife. We learned to avoid the front door, with the untrodden grass before it; for it led to the forbidden regions of the front parlor, open only for weddings and funerals, with its reed organ, its stiff haircloth furniture, its tinted family photographs, and its whatnot, laden with the particular treasure of the house and with the family album.

A little farther afield—about a mile and a half—lay the Shaker village. This was a particular treasure; a Protestant monastery, where the brothers and sisters of a sect doomed to perpetual celibacy sat on opposite sides of the aisle of their little chapel, dressed in an extreme version of the traditionally austere Quaker costume. I remember venerable Sister Elizabeth, and Sister Anne as well, who retained the worldly coquetry of wearing false hair under her coal-scuttle straw bonnet. One or the other would preside at the little shop in their great empty main building. They sold souvenirs and simples, as well as sugared orange peel and enormous disks of sugar flavored with peppermint and wintergreen. These were ridiculously inexpensive, and were the one sort of sweet which our parents allowed us to eat as far as our appetites might go.

The colony must have been about a century old and had an atmosphere of antiquity and permanence about it which was far more European than American. A celibate sect is always likely to find recruitment very difficult, and even though the Shakers would adopt children in the hope that they would grow up in their austere faith, something usually happened during adolescence or immediately after, and the youngsters almost always abjured the righteous faith of their foster parents to go the way of Satan and the flesh. Thus the great communal workshops and the two-story stone barns stood empty, and the fields went half tilled, while the outlying community houses under their spreading fir trees became orphanages or boardinghouses. The cemetery was a waste of weeds and brambles battening on the bones below, and the landing platforms which Shaker modesty had ordained to be built before each house, in order that the women mounting into the carriages should not display their feet in an unseemly and indecorous manner, were rotting away.

For some years Constance and I had been continually at odds with each other, and this seemed to my inexperienced parents a sign of original sin. I call them inexperienced because they were just beginning to realize the conflicts implicit in the growing up of brother and sister. Now, however, I was eight and Constance was four, and there began to be a possibility of our being comrades. I know that we explored the thirty-acre farm together, and that for the first time I began to see her as an individual.

Although my new country life had its delights, the breaking of my acquaintance with children of my own age was a strong countervailing disadvantage. I found, indeed, a group of children from Ayer and from neighboring farms with whom I could play. But the frequency of my play was cut down by our relative isolation. Indeed, I never again fully caught up with the richness of companionship which seemed to me on retrospect to belong to our years on Avon Street. I know very well that the disadvantages of being too much alone and isolated were not easy to overcome in view of the financial struggles of the family at that time, but the effects were serious and long-lasting. When I left Cambridge for Harvard, I broke up the acquaintanceships of my early childhood, and although I made new ones in Ayer and later in Medford, I was never again to feel the continuity of so rich an environment of childhood friendships.

« VII »

A CHILD AMONG ADOLESCENTS

Ayer High School, 1903–1906

FATHER intended to live on the farm in Harvard and commute into Cambridge each day. He was a very busy man, for, as I said, he had undertaken to translate twenty-four volumes of Tolstoy into English in a period of two years. This was a tremendous task to add to his Harvard teaching and the running of the farm. The amount of time he could devote to my education was limited, so he began looking around for a school to which he could send me. Even so, he intended to go over my lessons with me each night. I was by this time too far along in my studies to profit by any ordinary grade school, and the only solution seemed to be to send me to some one of the high schools in the neighborhood, and to let me find my own level. The Ayer high school was willing to try this unorthodox experiment. Ayer suited my father perfectly, as it was the nearest station on the main railroad to Boston and he had to drive there every morning to take the Cambridge train, leaving his horse and carriage in an Ayer livery stable till he should return in the evening.

A CHILD AMONG ADOLESCENTS

I entered the Ayer high school in the fall of 1903 at the age of nine as a special student. We left the problem of my eventual classification for the future to decide. It soon became clear that the greater part of my work belonged to the third year of high school, so when the year was over I was transferred to the senior class to be graduated in June, 1906.

The brains and conscience of the school were Miss Laura Leavitt, who has just retired after fifty years of service. She was gentle but firm, and an excellent classicist, with a feeling for Latin which went far beyond the perfunctory requirements of the average high school. I read Caesar and Cicero with her my first year, and Virgil my second. I also studied algebra and geometry, but these courses were largely review for me. I studied English literature and German from teachers who have left no particular impression on me. They were probably young women filling in their time between graduation from college and matrimony.

Although I could recite my lessons as well as most of the older pupils, and although my sight translations from the Latin were reasonably acceptable, I was socially an undeveloped child. I had not attended school since I had gone to the Peabody School in Cambridge at the age of eight, and I had never attended school regularly. Now at the Ayer high school the seats were much too big for me and my adolescent fellow students seemed to me already full adults. I know that Miss Leavitt tried to relieve me from the alarm of being in this unfamiliar place among unfamiliar figures, and on one occasion during my first few months at school, she took me on her lap during a recitation of the class. This kind act did not lead to any outburst of laughter or ridicule by the class, who seemed to consider me as the equivalent of their kid brothers. It was quite natural for a friendly teacher to take such a child on her lap when he visited the high school.

Of course, such treatment was in the long run incompatible

with the proper discipline of the school, and before long I had learned the elements of schoolroom behavior. The discrepancy between my age and that of my classmates continued to protect me from their ridicule. I think this would have been less true had I been only four years younger than they rather than seven. They viewed me socially as an eccentric child, not an underage adolescent. Hence, it was fortunate that the school shared a building with what would now be called a junior high school, where I was able to find some playmates among eleven- and twelve-year-olds, some of whom were the younger brothers of my classmates.

My training and social contacts at high school were only the obverse of the coin. The reverse was my continuing recitation to my father at home. My routine when I was in high school scarcely differed from that which had obtained when he was my full teacher. Whatever the school subject was, I had to recite it before him. He was busy with his translation of Tolstoy and could scarcely devote full attention to me even during my recitation hours. Thus I would come into the room and sit down before a father dashing off translations on the typewriter—an old Blickensderfer with an interchangeable type wheel that permitted my father to write in many languages—or immersed in the correction of endless ribbons of galley proof. I would recite my lessons to him with scarcely a sign that he was listening to me. And in fact he was listening with only half an ear. But half an ear was fully adequate to catch any mistake of mine, and there were always mistakes. My father had reproved me for these when I was a boy of seven or eight, and going to high school made no difference in the matter whatever. Although any success generally led at most to a perfunctory half-unconscious word of praise such as "All right," or "Very good, you can go and play now," failure was punished, if not by blows, by words that were not very far from blows.

A CHILD AMONG ADOLESCENTS

When I was dismissed from my responsibility with my father, I often spent the afternoon with Frank Brown. He was a boy of my age, the son of the local druggist and the nephew of Miss Leavitt, and he became a lifelong friend. We lived only two miles apart, so that it was not difficult for me to play with him after school or to walk out to see him on Saturdays and Sundays. His family looked favorably upon our acquaintanceship, and I have always considered them among my dearest friends.

Frank and I used to punt our boat from the pond up by the side of the old seventeenth-century mill dam into the brook and work it over stones and shallows until we came into a dark tunnel in the bushes, which led in a mile or two to a back road to Harvard Center. We would imagine all sorts of weird possibilities in an old slough in the woods. We poked sticks into it to see the bubbles of marsh gas rise and burst. We captured frogs and tadpoles in the pond and tried to make pets of these unappreciative and refractory animals.

Once I burned the skin off the back of Frank's hand when we tried to make firecrackers out of material snitched from his father's drugstore. On another occasion we loaded up a tire pump with water and lay in wait on the piazza to spray one of the early motorcars of that remote epoch. We assembled old cartons and carriage wheels into a ramshackle toy train and played railway man. And sometimes we went up to the attic, where we spent part of our time reading *Treasure Island* or *Black Beauty* and part of our time assembling bits of electrical apparatus to make an electrical bell. We once assembled something we conceived to be a wireless. We were boys as boys always have been and as boys will always be. Certainly I was neither particularly oppressed nor particularly impressed by my unusual school status at this age.

On one day every two weeks the high school had a debate and an oratorical contest in which the children recited standard passages from a compilation made for the purpose. Toward the middle of the vacation between my two years, I decided to write a philosophical paper which I could use for future occasions of this sort. I recited it the next winter, but not as an active participant in the competition. It was called the *Theory of Ignorance,* and was a philosophical demonstration of the incompleteness of all knowledge. Of course, the paper was unsuited for the purpose and beyond my age. But my father liked it, and as a reward for this paper he took a long trolley ride with me to spend a few days at Greenacre, Maine, near Portsmouth, New Hampshire, among the mists of the Piscataqua River. Greenacre was a colony of Bahaists, receptive to all forms of oriental religiosity. It represented a trend which now belongs rather to Los Angeles than to New England. I wonder what some of the good New England Bahaists must have thought when they found that Bahaism was a Sufiist variant of Islam.

Old Mill Farm was a real working farm, with cows, horses, and all the rest of it, under the supervision of a hired man and his wife. My personal possessions among the livestock of the farm were a goat and my special crony, my shepherd dog, Rex. Rex stayed with us until 1911 when my parents could no longer tolerate his habit of chasing automobiles, and they decided that he had better be put away. Though it may have been necessary, I could scarcely bring myself to consider it as anything but an act of treason to an old friend. The goat was bought by my parents as a present to me, to pull a little carriage which they had the hired man make. The carriage was good fun as a toy but rather unsatisfactory as a means of transportation. Rex and the goat seemed to have misunderstandings about a good many issues. The goat's horns and hard head were a perfectly adequate match for Rex's teeth.

The hardest time of the year was the late winter and early

A CHILD AMONG ADOLESCENTS

spring. Country roads were not paved at the time, and the carts and carriages made them into a mass of ruts which froze into brutally impassable ridges. A neighbor about a half mile away used to invite me to come over in this miserable, dreary time and play bezique with him and his wife. I had a great deal of time alone in the house to spend reading in father's library. I was particularly fascinated by Isaac Taylor's book on the alphabet, which I knew nearly from cover to cover.

But summer was different. Besides rowing and swimming in the pond, a little desultory botanizing, and mushroom-hunting trips with my father, I used to play with Homer and Tyler Rogers, two boys about my age who lived on a neighboring farm. We almost blew ourselves up by trying to make an internal combustion engine out of a tin fly spray, and we nearly shocked ourselves to death in amateur radio experiments, using apparatus which my father bought for me and which I was never able to put to any really effective use.

Father encouraged me to raise a garden, although he was not very enthusiastic over my horticultural proficiency. I dragged one load of beans after me in a boy's express wagon and managed to sell it to Mr. Donlan, the Ayer grocer. Donlan, who combined his grocery business with an agency of a steamship company, was a particular crony of my father's and they used to talk Gaelic together. To keep up with Mr. Donlan's Gaelic, Father borrowed a few books of Irish fairy stories from the Harvard library. He used to translate them to me as I lay in bed, and I was astonished at the grotesqueness and the formlessness of the tales, so different from the tales of Grimm to which I had been accustomed.

Late in the winter my father had a visit from Professor Milyukoff, a member of the Russian Duma (or limited parliament), an authority on political institutions and later a cabinet member under the ill-fated Kerensky regime. Milyukoff was a tall, genial, bearded Russian, and since he came about Christmas time, he brought my sister and me tufted children's snowshoes, with which we could traverse the white countryside as with seven-league boots. My parents had already their own snowshoes, and found that the regions that were too swampy to cross at other seasons were now as accessible as the open highway.

Milyukoff was writing a book about American political institutions, and Father guided him about the local points of interest in political and social history. Our neighbor Farmer Brown drove us around in his sleigh to the Shaker villages, to the estate of a neighboring single-taxer, and to the cottage of Fruitlands, where the Alcotts had lived after the breakdown of the Brook Farm project. Father explained everything to Milyukoff—or at least I thought he did, but I could not understand the Russian in which they spoke.

In the spring of 1906 when I was eleven years old, my brother Fritz was born. He was always a frail child, and I shall have more to say later concerning the problems of his development and education. My sister Bertha was seven years younger than I, but still near enough to share with me a considerable period of growth and development. But Fritz came at a time when I was already approaching adolescence; when he himself had come to adolescence I was already a young adult busy with the problems of making a career and of finding my social place in the world of ideas, so we could never be companions.

As I have said, I was put into the senior class at the beginning of my second year at the Ayer high school. I was almost eleven and full of rebellion. I had a wild idea (which never came to any sort of expression, even among my closest companions) of forming some sort of organization among children of my age to resist the authority of their elders. Then I had fits of conscience, and wondered if I had not committed

A CHILD AMONG ADOLESCENTS

a crime equivalent to treason by even contemplating such a thing. I consoled myself by thinking that even if I had, I was too young to be subject to serious punishment for it.

Toward the end of the high school spring term, I used to eat lunch with some of my schoolmates and teachers in a scrubby second-growth wild cherry grove near the school building. The ground was carpeted with anemones and violets, and there was an occasional lady's-slipper. The warm spring sun, shining through the branches which were as yet only covered with a green fuzz of foliage, summoned to a new life and activity.

There, in my last year of high school and at the age of eleven, I fell in love with a girl who played the piano at our school concerts. She was about fifteen, the freckled daughter of a railroad man. Futile as it was, it was real love, and not the almost sexless affection of undeveloped children. She was developed beyond her years. I was only eleven, but I was not even physically the typical eleven-year-old. My make-up was a mosaic of elements as young as eight and at least as old as fourteen. This calf love was quite as ridiculous to me as it must have seemed to others, and I was ashamed of it. I tried to show off in what was actually the least effective way open to me, and to compose a piece of music for her—I, the least musical of all boys. Like so many of these primitive attempts at composition, it sounded like nothing so much as the black keys of the piano struck in succession.

Of course, nothing could come of this friendship, not even an "affair." Besides the fact that I was a child in age, I was too alarmed at the new and half-understood powers within me to break with my infancy and indulge in unpermitted pleasures. My parents' inquiries of me and of other people showed them that this girl was not leading my body to destruction and my soul to perdition—though there had never been the slightest danger. The experience marked the end of carefree childhood. Little as I wished to grow up, I found myself rushing toward maturity, with its unknown duties and possibilities.

Calf love is the experience of every normal boy. But after a couple of years the boy finds himself associated with girls about his own age with whom he learns to be at home. And by the time he is in college, he is already sufficiently hopeful of eventual success to court girls seriously with a view to marriage not to be too long delayed. My calf love, however, was especially early, and when I entered my twenties I was still not nearly out of the woods, nor was I able to think of marriage.

The end of the school year was occupied with the graduation parties of my classmates who were then seventeen and eighteen years old. Even when I was the nominal host, and the guests arrived at Old Mill Farm in brakes rented from the Shakers, I was rather an outsider at the feast. I sat on one side of the room in the kneehole of the desk and watched the dancing as a ritual in which I had no part.

After the round of festivities and the graduation ceremonies, I passed the summer on Old Mill Farm with my *St. Nicholas Magazine* and my Ayer playmates and occasional visits to the farm of Homer and Tyler Rogers. I had tried several times to get a contribution accepted in the St. Nicholas League, that cradle of young artists, poets, and novelists, but the best I could make was honorable mention, and I rated that only once. I had to content myself with purchasable pleasures. I got hold of a cheap Brownie camera about this time; and I hoped to buy an air rifle, but because my parents frowned on this, the best I could get was a popgun with a cork.

I owe a great deal to my Ayer friends. I was given a chance to go through some of the gawkiest stages of growing up in an atmosphere of sympathy and understanding. In a larger school it might have been much harder to come by this understand-

A CHILD AMONG ADOLESCENTS

ing. My individuality and my privacy were respected by my teachers, my playmates, and my older schoolmates. I was treated with particular affection and understanding by Miss Leavitt. I had a chance to see the democracy of my country at its best, in the form in which it is embodied in the small New England town. I was prepared and ripened for the outer world, and for my college experiences.

Since leaving high school, I have visited Ayer several times although with long intervals between visits. I have seen the town burst from its status as a railway junction into that of a military cantonment, and I have seen the greater part of its lines of railway go away. I have seen the Second World War inflate the village once more, and I suppose that I shall see it sink again into relative unimportance. Yet, through all these vicissitudes, there has been a definite progress toward unity and serenity on the part of the families I have known. They are people living in a small town, but they are emphatically not small-town people. They are well read in an age of little reading. And they know the theater well, although the nearest theater is thirty-five miles away. Two generations have risen to maturity since I left the place, growing up in an atmosphere of love and of deference. I have the impression that my friends in this small industrial town represent a sort of stability without snobbishness which is universal rather than provincial, and that the structure of their society compares well with the best that a similar place in Europe would have to offer. When I go back among them it is expected of me, and rightly expected of me, that I revert in some measure to my status as a boy among the elders of the family. And I do so gratefully, with a sense of roots and security which is beyond price for me.

« VIII »

COLLEGE MAN IN SHORT TROUSERS

September, 1906–June, 1909

WITH MY HIGH SCHOOL days over, my father had decided to send me to Tufts College rather than to risk the strain of the Harvard entrance examinations and the inordinate publicity which could have been caused by sending an eleven-year-old boy to Harvard. Tufts was an excellent small college, so near to Harvard as to be overshadowed by it in the public eye. By virtue of that very fact, it shared in the intellectual advantages of the Boston metropolis. It was possible for us to live near Tufts on Medford Hillside, and for Father to take the trolley from there every day to his Harvard work.

I was admitted to Tufts on the basis of my high school record and a few easy examinations, which in my case were mostly oral. We bought a nearly finished house on the Hillside from the contractor builder, who lived next door to us, and had him complete it in accordance with our requirements.

We came down from the Old Mill Farm a little early to get settled in our new house and to familiarize ourselves with the college. I read the college catalogue assiduously, and was then more familiar with the details of Tufts College than at any time later.

I began to get acquainted with the children of the neighborhood. In my early reading I had learned something about hypnotism and decided to try it out myself. I succeeded in nothing, except in offending and terrifying the parents of my playmates. I played a good deal with children of my own age, but without any great community of interest. I found the clerk at the corner drugstore an interesting young medical student, who was prepared to discuss my scientific reading with me and who seemed to be acquainted with the whole of the writings of Herbert Spencer. I have since found Herbert Spencer to be one of the most colossal bores of the nineteenth century, but at that time I held him in esteem.

My duties began with the beginning of term. I was deeply impressed by the age and dignity of my professors, and I find it hard to be aware that I am now much older than most of them were then. I did not find it easy to make the transition between the special privileges of a child, which I had received in high school, and the more dignified relation I was now to have with these older men.

I started to study Greek under a rather remarkable professor by the name of Wade. His family had come from the neighborhood of Tufts College; and as a boy, while stealing a ride on a Boston-and-Maine freight car, he had fallen off and lost a leg. He must always have been shy, and this accident had made him a lonely man; but it had not seemed to interfere greatly with his love for European and Near Eastern travel. He would spend every summer abroad, and he seemed to know every relic of the classical world, whether it was a statue or a local tradition, from the Pillars of Hercules to Mesopotamia. He had a true poetic sympathy with the Greek classics as well as a gift of imparting this sympathy to others. His lantern lectures on Greek art were a delight to me. My father was fond of him, and he used to come over to our house. I would play on the floor and listen with fascination to the wide-ranging conversation between the two men. If anything could have made a classicist of me, it was these experiences.

I had not yet reached the proper stage of social maturity for my English courses. Moreover, the mere mechanics of writing were a serious hurdle to take. My mechanical clumsiness in writing tended to make me omit any word that I could eliminate, and to force me into a great crabbedness of style.

I was already beyond the normal freshman work in mathematics. There was no course which exactly fitted my requirements, so Professor Ransom took me on in a reading course on the Theory of Equations. Professor Ransom has only just retired from Tufts College after a half-century of service. He was a young man when I studied with him, and it stands to reason that he cannot be young now; but in the course of the years I have seen very little change in his vigorous, alert walk, in the forward poise of his bearded chin, and in his enthusiasms and interests. He was a zealot, but a self-effacing zealot. The course was really over my head, particularly in the parts concerning Galois' theory, but with a great deal of help from Professor Ransom I was able to get through. I had started my mathematics at the hard end. Never again at Tufts did I have a mathematics course that demanded so much of me.

I studied German under Professor Fay, known as "Tard" Fay from his lateness to class. He was a thoroughly cultured gentleman with a strong sense of literary values in French and German; and in addition, he was a great mountain climber. I believe that a mountain in the Canadian Rockies bears his name. Naturally, this seemed very romantic to me. Although we did a little bit of easy prose reading in German, the part of the course that attracted me the most was a collection of German lyrics. Here Professor Fay's efforts were more

COLLEGE MAN IN SHORT TROUSERS

than supplemented by the emotion with which my father would recite to me the many German poems which he knew by heart, and by the work of memorization which was both part of my class duties and which in its way was very welcome to me. My physics classes consisted as usual of recitations and some lectures and demonstrations. It took me some time to develop a proper physical sense to enable me to handle my exercises correctly, but I was always delighted in the demonstrations. I was equally delighted in my work in the chemical laboratory where, in my last year, I studied organic chemistry at probably the greatest cost in apparatus per experiment ever run up by any Tufts undergraduate.

I had a friend and neighbor, Eliot Quincy Adams, who was an undergraduate at M.I.T. during the time I was an undergraduate at Tufts. He introduced me to the possibility of representing four-dimensional figures on the plane or in three-space, and to the study of the four-dimensional regular figures. On one occasion we tried to make a water-dripping electrical machine from old tin cans.

I had several other extracurricular adventures in physics and engineering, more especially in the study of electricity. I shared electrical experiments with a Medford neighbor. We used to generate electricity by turning a hand-run dynamo for the making of colloidal gold and colloidal silver. Whether we actually made these substances I cannot remember, but we thought we did. We also made attempts to realize in practice two physical ideas of mine. One of them was an electromagnetic coherer for radio messages different from the electrostatic coherer of Branly. It depended on the effect of a magnetic field independently of its direction, in compressing a mass of iron filings and powdered carbon, and thus in changing its resistance. There were times when we thought we had obtained a positive effect, but we were not certain whether it was due to this magnetic cohesion or to something quite different. Nevertheless, the idea was sound and if the day of all such devices had not passed with the invention of the vacuum tube, I should be interested to undertake these experiments over again from the beginning.

The other piece of apparatus we tried out was an electrostatic transformer. It depended on the fact that the energy or charge of a condenser is carried as a dielectric strain. The trick was to charge a rotating glass disk or series of disks through electrodes arranged in parallel and to discharge them through electrodes arranged in series. It differed from the electromagnetic transformer in acting on direct currents, and also in the fact that it was essential to the apparatus that the disks should be revolved. We broke an indefinite number of panes of glass in trying to make the machine, and we never quite got it to work. Unbeknown to us, the idea was already in the literature and had been there for a long time. In fact, I have seen a very similar piece of apparatus within the last two years in the laboratories of the School of Engineering of the University of Mexico. It functioned very well. Two successive stages of this machine multiplied the potential by several thousand.

I had an early interest in wireless. I believe that on rare occasions I was able to get a few successive dots and dashes of code signal from the wireless apparatus which I had on my study desk. I was able neither to learn code nor to distinguish myself as a practical constructor of radio sets.

I was socially dependent far more on those of about my age than on the college students with whom I was to study. I was a child of eleven when I entered and was in short trousers at that. My life was sharply divided between the sphere of the student and that of the child.

I was not so much a mixture of child and man as wholly a child for purposes of companionship and nearly completely a man for purposes of study. Both my playmates and the college students were aware of this. My playmates accepted me as a child with them, although I might have been a slightly incomprehensible child, while my fellow students were willing to allow me to participate in their bull sessions if I wasn't too loud and too insistent. I was homesick for the earlier days when I had had a wealth of playmates in Cambridge.

While I was at Tufts, I continued to spend summers on Old Mill Farm where I kept up my contacts with my Ayer friends, and where an occasional Tufts fellow student would come up to visit us. All summers were alike, with the same mushroom-hunting trips, the same general botanizing, the same tramps, and the same swimming in the pond. With my increasing maturity, I came to be taken in as a part of the family on those occasions when we had visits from my parents' friends.

In term time I tried to renew my acquaintance with my old friends of Avon Street. These attempts at renewing the past were not uniformly successful, and finally ceased altogether. Medford Hillside was too far from Cambridge to make it easy for me to visit my friends except on week ends. In addition, with my own isolation from Avon Street, I had become more exacting and jealous. What is more, my Avon Street comrades were growing in divergent directions. The King boys were showing, indeed, an increasing interest in science. While I later came to see a good deal of the King boys again and to play with them in a desultory fashion in their basement laboratory, I saw very little more of many other Cambridge acquaintances.

I have already had something to say about my taste in scientific reading. My nonscientific reading was omnivorous. I made free use of the resources of the various public libraries to which I had access, and I spent a great deal of time in the Children's Room at the Boston Public Library.

That I loved the writings of Jules Verne I have said already, and for adventure I alternated them with Cooper and Mayne

Reid. Later, when I came to years of greater discretion and was able to stand a stronger literary diet, Hugo and Dumas were added to my repertory. Dumas in particular was a writer whom I could not bear to put down, and I have spent many hours unaware of the world about me, immersed in the adventures of D'Artagnan or the Count of Monte Cristo.

Naturally I read many of the children's books which the public library had hoarded from an older generation. Louisa Alcott was pleasant enough, but I was a young male snob, and I considered that it was mostly for girls. Horatio Alger combined a superficial appearance of prudence and morality with a crassness of standards of success which was very distasteful to me. I even ventured on a course of reading dime novels, but found them pretty thin. My favorite author among the American boys' writers in the narrower sense was J. Trowbridge, even though his tales of New England and upper New York State boyhood do not impress me quite as much as they once did. On the other hand, I think his three Civil War novels, *Cudjo's Cave, The Drummer Boy,* and *Three Scouts,* represent as high a level as can be achieved in boys' war stories.

I used to buy the old *Strand Magazine* at the newsstands. This was an English periodical which did very well in the United States for many years. It contained some of the Sherlock Holmes stories, some excellent children's tales by Evelyn Nesbit, and a few of A. E. W. Mason's excellent detective stories. It was better written than the majority of American periodicals at the time, and introduced me to many new authors and renewed in my mind much of the quaint, grim aspect of London.

I was not an indoor boy even in the winter. The road that passed by the Tufts College Reservoir offered a great opportunity for coasting. I also enjoyed the crisp, bitter air of winter as I drew it into my lungs on my jaunts with a friend on his newspaper route; and there was a thrill in rushing from building to building on the open spaces of College Hill even in the intense and numbing cold.

My father took a certain glibness of vocabulary of philosophy on my part as indicating that it was my true intellectual field and he encouraged me in it. Therefore in my second year at Tufts, I took several courses in philosophy and psychology under Professor Cushman. He was rather an amateur in philosophy.

The two philosophers who influenced me most in my reading were Spinoza and Leibnitz. The pantheism of Spinoza and the pseudomathematical language of his ethics mask the fact that his is one of the greatest religious books of history; and if it is read consecutively instead of broken up into axioms and theorems, it represents a magnificent exaltation of style and an exertion of human dignity as well as of the dignity of the universe. As for Leibnitz, I have never been able to reconcile my admiration for him as the last great universal genius of philosophy with my contempt for him as a courtier, a place-seeker, and a snob.

The rather dilute material of my courses in philosophy and psychology showed up poorly in comparison with the outside reading and in particular with the great books of Professor William James, which I devoured almost as much as literary tidbits as for their serious content. I learned that James was one of the heroes of my father, and it was not long before I had a chance to visit him in his own house. I do not recall the visit too distinctly, but I have the impression of an amiable, elderly, bearded man who was kind to me in my confusion, and who later invited me to sit in on his Lowell lectures on pragmatism. I attended them and was delighted when Professor James presented my father with a copy of the book in which he embodied this series of lectures. I learned later that the book was really intended for me, and that neither James nor my father wished to raise my conceit by having him make the gift directly.

I do not have the impression that James was at his best in pragmatism. In the more concrete material of psychology, his insight showed itself in every paragraph; but pure logic was never his strongest point. It is one of the clichés in American intellectual history that while Henry James wrote novels in the style of a philosopher, his brother William James wrote philosophy in the style of a novelist. There was more than the style of the novelist in William James, and perhaps less of the philosopher than one might have thought, for his ability to evoke the concrete was to my mind many times greater than his ability to organize it in a cogent logical form.

Also during my second year at Tufts, I found the biology museum and laboratory a place of fascination. The custodian of the live animal house, who was janitor as well, became one of my particular cronies. These noncommissioned officers of science, without whom no laboratory could ever function, are a fascinating group of people, and particularly attract the imagination of a young boy with ambition to go into science. I thought that I would take a fling at a little biological work. I had already gone with Professor Lambert and a group of students on several spring biological excursions to the Middlesex Falls and elsewhere and watched them collect frog spawn, algae, and other objects of biological interest.

I had long shown an interest in biological matters, and my father wished to find out whether it would be worth my while to take up biology for my specialty in my further studies. We took a train trip together to Wood's Hole, where Professor Parker of the Harvard Biology Department allowed me to try my hand in dissecting some dogfish. All that I remember is that my dissections were not particularly brilliant and that in a few days a notice appeared on the dock where I was working saying: "No fish cut up here."

COLLEGE MAN IN SHORT TROUSERS

In my last year I decided to have a serious try in biology. I took Kingsley's course on the comparative anatomy of vertebrates. Kingsley, by the way, was the author of the *Natural History* which had so intrigued me about my eighth year. He was a small, birdlike, alert man, and the most inspiring scientist whom I met in my undergraduate days. I had no trouble whatsoever with the class work, for I had always had a good sense of the arrangement of things; but my dissection was too fast and too sloppy. Kingsley saw to it that I had enough work to do, and he gave me a large number of reptilian, amphibian, and mammalian skulls to learn in order to see whether I could discover the secret of their homology. Even here I worked too fast and too messily. I used to spend a great deal of time in the library of the laboratory where I would read such books as Bateson's *Material for the Study of Variation*.

Now biological study may have a morbid attraction for a young student. His legitimate curiosity is mixed with a prurient interest in the painful and the disgusting. I was aware of this confusion in my own motives. I have said that there were passages in my books, scientific treatises and fairy tales alike, which I turned over quickly to avoid but which I would seize on now and then with grim pleasure. The humanitarian tracts about antivivisection and vegetarianism which cluttered up our study table increased my confusion by their exaggeration. In this confusion, I found myself in more than one doubtful situation.

The most serious experience of this confusion occurred in my last year at Tufts. Several of us had been in the habit of doing our dissection of the cat with the aid of a human anatomy—I forget now whether it was Quain's or Gray's. This was a highly desirable practice inasmuch as the anatomy of the cat and of man, though closely parallel, are not literally identical, and so the very differences put us on our mettle and made us better observers. Now some of these human anatomies contain interesting observations concerning the ligature of arteries and the new anastomoses which are established to set the circulation up again. Two or three of us were particularly interested in these passages. The older boys were naturally much more mature than I was, but I am afraid that I must admit that I was the ringleader in the affair. Through the complacency of the janitor we obtained a guinea pig and ligated one of the femoral arteries. I do not remember whether we used anesthetic or not, although it is my faint impression that we did after a fashion, and that the animal got a whiff of ether. The surgery was botched, as we had not properly separated the artery from the accompanying vein and nerve, and the animal died. When Professor Kingsley found out about our misadventure, he was very indignant as the vivisection was undoubtedly a criminal undertaking and might very well have forfeited important privileges of the laboratory. Although I was not effectively punished from outside, I was humiliated and deeply disturbed. It was very clear to me that I could give to myself no account of my motives that would stand inspection before the court of my own conscience. I tried to bury the act in a premature obliviscence which, of course, drove it further into my consciousness. My feeling of guilt about this episode led me to greater tension.

In spite of this interest in biology, it was in mathematics that I was graduated. I had studied mathematics every year in college, largely under Dean Wren whose point of view was more nearly that of the engineer than that of Professor Ransom, who had taught me in my freshman year. I found the courses on calculus and differential equations quite easy, and I used to discuss them with my father who was thoroughly oriented in the ordinary college mathematics. For my routine of double recitation had not changed so far as my mathematical and cultural courses were concerned. In these my father remained my complete master, and there was not the slightest slackening in his stream of invective.

I was graduated in the spring of 1909, having completed my academic course in three years. This does not represent quite as much of a triumph as it seems, for I had fewer distractions than other boys. Only the child can devote his whole life to uninterrupted study.

I decided to go to the Harvard Graduate School the next year to continue my work in zoology. This was primarily my decision but Father was rather unwilling to concur in it. He had thought it might be possible for me to go to medical school, but Professor Walter B. Cannon advised him strongly against it, saying that my youth would be an even greater handicap there than elsewhere.

Since I was no longer to be at Tufts, my father planned to move into Cambridge for the next year. This meant the buying or building of a house. The persuasion of a firm of architect colleagues at Harvard led my father to buy a pair of lots on the corner of Hubbard Park and Sparks Street, on one of which was erected a truly magnificent structure symbolic of the family's growing prosperity. This involved complicated maneuvers, aimed at disposing of the house on Medford Hillside and of the Old Mill Farm at Harvard. It became a part of the family articles of faith to consider that these maneuvers showed an extraordinary foresight and knowingness. The extra lot was to have been sold as soon as we could find a purchaser, but no purchaser presented himself until the house and lot were sold together fifteen years later.

At any rate, we could not spend the summer at Old Mill Farm. We returned to Harvard for the first part of the summer, but in a different part of the township and in an old, tumbledown house which proved to be bad for the health of all of us. We gave up the house before the end of the summer, and left Harvard to finish the vacation in a boardinghouse at

Winthrop, from which location my father could watch the completion of our new house. Through the kind offices of two ladies who worked at the Harvard library, we found a tolerable boardinghouse. I had time to waste until Harvard should open. I spent it, partly in the Winthrop public library, partly in Boston where I visited the museums and movies, and partly among the mechanical amusements of Revere Beach.

This was the exciting time of the discovery of the North Pole and the conflicting reports which came in from Cook and Peary. I remember Cook's ingratiating newspaper personality, and the hopes we had falsely put on him. The Mutt and Jeff caricatures of Bud Fisher had but recently started, and they concerned themselves largely with the tragicomedy of polar exploration. As they were portrayed then, Mutt and Jeff could have scarcely been less than thirty years old. It is astounding how spry they remain at the age of more than seventy-two.

One scientific idea had haunted me that summer. It was that the vertebrate embryo represented a coelenterate polyp in which the pouches leading into the arms had become the myotomes. The nerve-belt about the mouth seemed to me to correspond to the brain and the spinal cord, and the bottom part of the central cavity to the vertebrate digestive tube. I remember that I used the microscope of our doctor in Ayer to examine some slides sent me by a friend from Wood's Hole, and I pestered the Carnegie Foundation with the request to allow me to do research on the subject. Of course, this came to nothing.

«IX»

NEITHER CHILD NOR YOUTH

I HAD NOT realized until my graduation how much the three years at Tufts had taken out of me. I was exhausted, but I could not stop the wheels from going around, and I could not rest.

I did not prosper physically that summer. Every time I got a scratch I festered mildly, and I was in a continual low fever. My emotional state corresponded with my physical condition. The feeling of growing up out of the protection of childhood into that of responsibility had not been welcome to me. With my undergraduate days over, and an unknown future confronting me, I felt at loose ends.

I had had my due share of the brief gratification of Commencement; but behind this happy moment were running the great questions: what should I do in the future and what hopes might I have of success?

The first question had been partly answered by my decision to do graduate work at Harvard. But the question of my success had an added poignancy. Though I had graduated *cum laude,* I had not been elected to Phi Beta Kappa. My record could be interpreted in two ways, and both my appointment and my nonappointment were defensible. But I was given to understand that the chief reason I did not receive the appointment was the doubt as to whether the future of an infant prodigy would justify the honor. This was the first time that I became fully aware of the fact that I was considered a freak of nature, and I began to suspect that some of those about me might be awaiting my failure.

Fifteen years later when I received the honor denied me at my graduation, I had begun to make my mark on the scientific world. To appoint me then was to bet on a horse after the race was over. The appointment at the time of my graduation would have meant a trust in myself and in my future which would have been a source of strength. For a good measure of conceit mingled in me with a greater measure of unsureness.

I have indeed taken a very dim view of all honor societies. This is, of course, the result of my own experience at Tufts, but it has been fortified and strengthened by my subsequent contact with such societies. The fundamental difficulty is that the recognition given by such societies—and indeed by universities in the awarding of honorary degrees—is secondary. They do not seek out young men deserving recognition, but award recognition largely on a basis of past recognition. There is thus a pyramid of honors for those who already have honors, and, per contra, an undervaluing of those who have behind them accomplishments rather than prior recognition.

There is a certain moral duty which I have felt here. I was only too conscious of the bridge that I had to cross to achieve any recognition whatever, and I resented the closed and serried ranks of my seniors as a direct obstacle to advancement and to confidence in myself. Therefore, when later recognition has come to me I have felt loath to be the beneficiary of a process of secondary recognition which I had resented as a young man. In this way, my early rejection by Phi Beta Kappa has strengthened me in a policy on the basis of which I have resigned from the National Academy of Sciences, and have discouraged my friends in the attempt to obtain for me similar honors elsewhere. I have not been absolutely consistent in this matter because there are cases in which the refusal of honors is viewed, not as a sturdy independence on the part of the recipient, but as an ungraciousness to worthy intellectual groups which are consciously or unconsciously seeking the support of their name. Be this as it may, my reaction is essentially the same at the present day as it has been for nearly forty years—that academic honors are essentially bad, and that other things being equal, I choose to avoid them.

Thus my graduation from Tufts forced me to face one of the greatest realizations that the infant prodigy must make: he is not wanted by the community. He has received no special rebuffs from his contemporaries. All children quarrel, and it is not until they reach the years of discretion that they acquire something superior to the social mores of the zoo. But as the infant prodigy comes to realize that the elders of the community are suspicious of him, he begins to fear reflections of this suspicion in the attitude of his contemporaries.

There is a tradition, not confined to the United States, that the child who makes an early start is intellectually drawing on his life capital of energy and is doomed to an early collapse and a permanent second-rateness, if not to the breadline and the madhouse.

My experience leads me to believe that the prodigy is desperately unsure of himself and underrates himself. Every child, in gaining emotional security, believes in the values of the world around him and thus starts by being, not a revolutionary, but an utter conservative. He wishes to believe that his elders, on whom he is dependent for the arrangement and control of the world in which he lives, are all wise and good.

When he discovers that they are not, he faces the necessity of loneliness and of forming his own judgment of a world that he can no longer fully trust. The prodigy shares this experience with every child, but added to it is the suffering which grows from belonging half to the adult world and half to the world of the children about him. Hence, he goes through a stage when his mass of conflicts is greater than that of most other children, and he is rarely a pretty picture.

In my earlier years, I had not been aware that I was an infant prodigy. Even in my high school days the matter just began to become clear to me, and during my college training I could not avoid facing it. One of the less agreeable consequences was that I was pestered by a swarm of reporters who were eager to sell my birthright at a penny a line. I soon learned that whining tone in which a reporter, bent on intruding on one's privacy, will tell you, "But my job depends on my getting this interview!" Finally I learned that reporters were on the whole to be avoided, and I ultimately developed enough fleetness of foot and ability at dodging to conduct a reporter across the college campus, later through the back alleys of Harvard Square, without giving his partner a chance to take a usable photograph.

Most of these articles appeared in the Sunday supplements of the newspapers. They belonged to a class of ephemeral literature which has long since returned to the gutters from which it emanated. While they flattered my childish desire for attention, both my parents and I recognized this for the sickly narcissism which it was; and it was not pleasant to see oneself endowed with a ten days' immortality between an account of a two-headed calf and the more-or-less true tale of the amours of the Count of X with the elderly wife of the millionaire Y.

It was the more serious articles that did the most harm. The suave and flattering articles of H. Addington Bruce[1] gave my father a chance to sound off with his unflattering theories of my education, while an occasional article in the trade journals of the educationalists[2] displayed to me the full tally of my gaucherie and social rejection.

I fear that Father himself was not immune to the temptation to grant interviews to the slicks about me and my training. In these interviews, he emphasized that I was essentially an average boy who had had the advantage of superlative training. I suppose that this was in part to prevent me from being conceited, and that it was no more than a half-representation of my father's true belief. Nevertheless, it rendered me more diffident as to my own ability than I would have otherwise been even under my father's scolding. In short, I had the worst of both worlds.

Besides the direct damage to me, these articles could only have accentuated that feeling of isolation forced on the prodigy by the hostility latent in the community around him.

The end of my college career forced me to take stock of myself and of my position in the world about me. In my exhausted state, this assessment and evaluation took on a dismal tone. I had for the first time become acutely conscious of the fact of death. I would count the fourteen and a half years that I had lived, and would measure my probable future life, and what I might expect of that. I could not read a novel without figuring out the ages of the characters and how many years they had left them to live, nor could I avoid looking up the lives of the authors of great books, and finding out for myself their own ages at the writing of these works and how many more years had been granted them. This obsession, of

[1] *Loc. cit.*
[2] Katherine Dolbear, ''Precocious Children,'' *Pedagogical Seminary,* Vol. 19, p. 463.

course, touched my relations to my parents and my grandparents, and for a period made my life utterly intolerable.

The fear of death ran parallel to and was reinforced by the fear of sin. My adventure in vivisection had brought a terrible awareness of possibilities in me of cruelty and the delight in cruelty which could only be brought into full focus with pain and suffering. My years in college had come at almost precisely the period in which I was turning from a boy into a very young and inexperienced man. The consciousness of the possibilities of manhood in me without any corresponding experience and worldly wisdom by which I could guide them led me into a frightful panic in which at times I sought for the blocked way back into childhood. I had been brought up in a doubly Puritan environment which supplemented the original intrinsic puritanism of the Jew with the puritanism of the New Englander. And the most elementary phases of my self-esteem from childhood into adolescence appeared to me either sinful or fraught with the possibilities of sin. These possibilities were things that I could not discuss freely even with my parents. My father would have taken what was on the whole a sympathetic attitude, but an attitude expressing a sympathy which was loath to go into details, and which on the whole was not likely to answer my inarticulate attempts to express myself about disagreeable matters with a willingness to listen and to find out what was really troubling me. On the other side, my mother combined a literal acceptance of the minutest code of puritanism with an adamantine unwillingness to admit that anything that a child of hers might do could be a possible offense against this code.

In other words, in these matters which were of great concern to me, I was met not so much with an accusing hostility as with a blockage of communication which left me substantially alone with my problems. This is no peculiar burden of the infant prodigy; it is the common lot of many adolescents, perhaps most. Yet when it is combined with the manifold other burdens that are thrust upon the prodigy, it is naturally accentuated.

If a child or a grandchild of mine should be as disturbed as I was, I should take him to a psychoanalyst, not with confidence that the treatment would be successful in some definitive way, but at least with the hope that there might be a certain understanding and a certain measure of relief. But in 1909 there were no psychoanalysts in America; or at least, if one or two venturesome disciples of Freud might have strayed so far, they were isolated practitioners. There was no tradition of resorting to them, and they were scarcely available to persons of the moderate means of a college professor. Moreover, even twenty years later, it would have seemed to my parents a blasphemy and a confession of defeat to admit even to themselves that a member of the family might need such treatment.

However, these remarks are after the fact. The fact is that it was then possible to allay neither the agonies of leaving childhood nor the sense of guilt almost inseparable from adolescent sexuality. Like many other adolescents, I walked in a dark tunnel of which I could not see the issue, nor did I know whether there was any. I did not emerge from this tunnel until I was nearly nineteen years old and had begun my studies at Cambridge University. My depression of the summer of 1909 did not suddenly end; rather it petered out.

My relationship with my father was gradually undergoing a change which I was only dimly aware of. Tufts had given me a partial release from his hitherto unrelaxing authority, for I tried my wings in such fields as biology where he could no longer lead me, and where I could hope to excel him. Indeed, my study of mathematics had won his approval, and at the same time had provided a field for me into whose later development it was impossible for him to follow me and break my independence. My study of mathematics gave me a consciousness of strength in a difficult field, and this was one of its great attractions. My mathematical ability at this time was a sword with which I could storm the gates of success. This was not a pretty or a moral attitude, but it was real, and it was justifiable.

From the time that I started to Tufts, my father used to tell me of the philological work that he was doing. Some of it concerned the early history of the Gypsies; some of it was devoted to moot points of philology, such as the origin of the Italian word *andare* and of the French word *aller*, some of it to Hecate worship and its influence in medieval Europe, some of it to Arabic influences on European languages. There was also a great deal of interesting research on the grand outlines of the relations between the established groups of languages such as the Indo-European, the Semitic, the Dravidian, et cetera; and later this comparative study was extended to the languages of the Americas and to the problem of African influence on these languages.

In all these investigations, my father combined a rare extent of linguistic information, an almost unexampled philological-historical sense, and a very modern mistrust of purely formal phonetic philology in favor of the more historical and empirical point of view which has come into ascendancy in the present generation. Father was a great admirer of Jespersen and his work; and when the time shall come to make an objective and judicial inquiry into the sources of modern philological ideas, I myself have no doubt that his name will stand beside that of Jespersen among the greatest in linguistic science.

Nevertheless, Father's intuition, although it was supported by an almost superhuman industry of reading and research, worked too fast for his formal logical processes. Philology was for him a piece of deductive work, a magnificent crossword puzzle; but I am afraid that he often gave up the writing of the words when he had still about a quarter of the paper left to cover. To a very considerable extent he knew what the remaining words were, but was simply too contemptuous of the powers of his readers to dot every *i* and cross every *t*. Nevertheless, I am convinced that (in a not appreciable and important fraction of the cases) he was not sure in his own mind of the last refinement of his work.

Now Father had come to philology with a thorough linguistic training, but without the sponsorship of any of the craftsmen of that guild. He was a man who liked popularity and acclaim, but he was essentially a lonely figure. The great foreign names in philology regarded him as an intruder, and his very ability made him at once dangerous and unpopular. At Harvard, the Slavic department formed a little autonomous enclave in the Division of Modern Languages, and although Father was consulted by everyone, he belonged to no one. Concealed behind his personal popularity, which was great, there was the inevitable distrust of the plodder for the genius. We all know that German aphorism credited to Ludwig Börne: when Pythagoras discovered the theorem of the right triangle, he sacrificed a hundred oxen; since then, whenever a new truth is unveiled, all oxen tremble.

Under these circumstances—a hermit in the midst of a great city—it was not unnatural that my father should turn to me for intellectual companionship and support. I was profoundly interested in his work, but I was not always convinced by every detail of his argument. When I showed my little faith by asking a question, Father was indignant. It was treason for me to question his slightest word. Indeed, I was not entitled to an opinion on most of the questions which he put to me, but I could only answer by the light of what little judgment was in me, and Father's whole precept and example were against saying yes where I could give no real assent. I should have realized, of course, that in much of Father's conversation with me I was merely a lay figure to represent the scientific public in a dialogue which Father was merely holding with his own doubts. Nevertheless, and despite all my protests that I did not know enough of the subject of which he was talking to give a reasonable opinion, my father would often demand of me a direct answer to a specific question. Then I might well find myself between a lying assent and an unwilling defiance. I preferred the defiance; but my father would certainly have seen through the falsity of the assent and would have berated me for my half-heartedness. It was unfair and I knew it was unfair; but I also knew that my father brought these issues to me out of a compelling inner necessity and that he was not a happy man.

« X »

THE SQUARE PEG

Harvard, 1909–1910

IN THE FALL of 1909 President Eliot retired from Harvard and was succeeded by Abbott Lawrence Lowell. I went from Winthrop to view this ceremony at which several honorary degrees were conferred in the open air in front of University Hall, and I enjoyed the academic pageantry.

I was not aware, and indeed few of us were at the time, that the departure of Eliot coincided with the end of a great age and the beginning of an age of smaller things. Eliot may have had the temperamental limitations of a New Englander, but he had the outlook of a scholar and man of the world. Lowell was dedicated to a Harvard that should be the private preserve of a ruling class.

The reign of Lowell soon began to contribute much to the prosperity of the members of the faculty, and my father was among those who had good cause to be financially grateful to the president. Yet this gratitude could never be untainted. If Lowell made the professors rich, it was because he wished them to be allies of the rich. He wished them to eschew the companionship of the common man, and to find their comrades in the circles of big business and big industry.

My parents were not originally aware that the cornucopia extended to the faculty by President Lowell had a razor blade concealed in it. Many years later when my father received at his retirement a letter nearly as peremptory as that which one gives to an unsatisfactory kitchen maid, President Lowell, this paragon of all virtues, became in the eyes of my father and mother a monster of iniquity. But he was neither a paragon nor a monster. He was essentially a superficially polished and rather ordinary man of conformist loyalties, who was loyal to his social class and to very little else.

In the early days of his presidency I used to attend the rather stiff and formal parties for college students at the President's House on Quincy Street. We learned to balance tea cups on our knees, and to hear tales of Mrs. Lowell's memories of the great frost that had tied up Boston harbor one remote winter in the past. The president would utter *obiter dicta* running from one end of the field of government and administration to the other, and would trot out his favorite idea, that while the government should make use of experts, they should be heard and not seen. He sang the praises of the amateur in politics, of the man who could judge everything well but who did not need to carry in his mind any particular content of information.

Term at Harvard began a few days before we moved into our new house in Cambridge. I was nearly fifteen years old, and I had decided to make my try for the doctor's degree in biology.

My first memories of my Harvard work are of collecting for my histology course a number of leeches from a small pond on the Fresh Pond reservation in Cambridge. My histology course began as a muddle and continued as a failure. I had neither the manual skill for the fine manipulation of delicate tissues nor the sense of order which is necessary for the proper performance of any intricate routine. I broke glass, bungled my section cutting, and could not follow the meticulous order of killing and fixing, staining, soaking and sectioning, which a competent histologist must master. I became a nuisance to my classmates and to myself.

My clumsiness and ineptness were due to a mixture of several factors. Probably there was a considerable amount of primary muscular lack of dexterity in my make-up, but this was not the whole show by any means. My eyes were another major factor; and though my eyesight was well corrected by efficient glasses and was at that time almost untiring, there are certain secondary disadvantages of the myope which are not immediately obvious to the average man. Muscular dexterity is neither a completely muscular nor a completely visual matter. It depends on the whole chain which starts in the eye, goes through to muscular action, and there continues in the scanning by the eye of the results of this muscular action. It is not only necessary for the muscular arc and the visual arc to be perfect, each by itself, but it is equally necessary that the relations between the two be precise and constant.

Now a boy wearing thick glasses has the visual images displaced through a considerable angle by a small displacement in the position of the glasses upon the nose. This means that the relation between visual position and muscular position is subject to a continual readjustment, and anything like an absolute correlation between these is not possible. Here we have a source of clumsiness which is important but which is not immediately obvious.

A further source of my awkwardness was psychic rather than physical. I was socially not yet adjusted to my environment and I was often inconsiderate, largely through an insufficient awareness of the exact consequences of my action. For example, I asked the other boys the time instead of getting a watch myself, and finally they gave me one. It is a little easier to understand this inconsiderateness if one realizes that although my personal expenses were taken care of very generously by my parents, I had not become accustomed to the complete control of any regular weekly allowance at all.

A further psychic hurdle which I had to overcome was impatience. This impatience was largely the result of a combination of my mental quickness and physical slowness. I would see the end to be accomplished long before I could labor through the manipulative stages that were to bring me there. When scientific work consists in meticulously careful and precise manipulation which is always to be accompanied by a neat record of progress, both written and graphical, impatience is a very real handicap. How much of a handicap this syndrome of clumsiness was I could not know until I had tried. I had moved into biology, not because it corresponded with what I knew I could do, but because it corresponded with what I wanted to do.

It was inevitable that those about me discouraged me from further work in zoology and all other sciences of experiment and observation. Nevertheless, I have subsequently done effective work together with physiologists and other laboratory scientists who are better experimenters than I, and I have made some definite contributions to modern physiological work.

There are many ways of being a scientist. All science originates in observation and experiment, and it is true that no man can achieve success who does not understand the fundamental methods and mores of observation and experiment. But it is not absolutely necessary to be a good observer with one's own eyes or to be a good experimenter with one's own hands. There is much more to observation and experiment than the mere collection of data. These data must be organized into a logical structure, and the experiments and observations by

THE SQUARE PEG

which they are obtained must be so framed that they will represent an adequate way of questioning nature.

The ideal scientist is without doubt the man who can both frame the question and carry out the questioning. There is no scarcity of those who can carry out with the utmost efficiency a program of this sort, even though they may lack the perspicacity to frame it: there are more good hands in science than there are good brains to direct them. Thus, although the clumsy, careless scientist is not the type to do the greater part of the work of science, there is other work for him in science if he is a man of understanding and good judgment.

It is not very difficult to recognize the all-round scientist of whose calling there is no doubt. It is the mark of the good teacher to recognize both the laboratory man who may do splendid work carrying out the strategies of others, and the manually clumsy intellectual whose ideas may be a guide and help to the former. When I was a graduate student at Harvard, my teachers did not recognize that despite all my grievous faults, I might still have a contribution to make to biology.

The late Professor Ronald Thaxter was, however, an exception. I took his course in cryptogamic botany. The lectures were elaborate and detailed accounts of the anatomy and phylogeny of algae and fungi, mosses, and the ferns and their allies. This normally involved the student in the taking of elaborate notes and the copying of diagrams which the professor had drawn on the board. The laboratory work consisted of the inspection and the drawing of living plants and sections of cryptogamic tissues. My laboratory work was worse than terrible: it was hopeless. Yet I got a B-plus in the course without having retained in my possession a single note.

Taking notes is the student's craft, and I did not master it. I think that there is a definite conflict between the use of a student's attention for the writing down of the best possible set of notes, and the use of his attention for the understanding of the speaker as he proceeds. The student must choose to do one or the other, and each has its advantages. If, like me, he is so clumsy of eye and hand that his notes are bound to be incomplete and only partly legible, he falls between two stools. If he decides to take notes at all, he has already destroyed much of his ability to grasp the argument in flight, and at the end of the course has nothing but a mass of illegible scribbles. Yet if his memory is as good as mine is, it is far better to give up the idea of taking notes and to organize in his mind the material as it comes to him from the speaker.

My comparative anatomy course went along in a manner intermediate between that of my histology course and that of my course in botany. My drawings were execrable, but I had a good understanding of the facts. My great tendency, as it had been with Kingsley, was to hurry the work rather than go into the fine points. This, indeed, has been my tendency throughout life. It is not difficult for me to explain. I have a rather quick insight into ideas, and an extreme lack of manual dexterity. It has always been difficult for me to hurry my manual efforts into a pace corresponding to the flow of my ideas, or to marshal my ideas into a sequence sufficiently slow to conform to the demands of my physical resources.

I found my physical recreation in the gymnasium. I joined a calisthenics class in which the usual setting-up exercises were combined with some mild folk dancing. I tried to join in the scrub basketball games that were being played in the basement of the gymnasium, but glasses were not to be thought of in the rough and tumble of the sport, and without glasses I was wholly useless.

I spent many happy hours in the library of the Harvard Union. It had recently been founded by the benevolent authorities as a "*club des sans club*" in protest against the exclusiveness of the Harvard clubs. It did not thrive under the new

Lowell regime. A new anti-Semitism was the order of the day, and the authorities began to think in terms of a *numerus clausus*. Whispers gradually appeared out of nowhere to the effect that the Union had become the headquarters of Jews and similar undesirables. These rumors received a sympathetic ear in my home. My mother questioned me about the Jewish character of the Union, and began to indicate to me that it might be a little better if I were not seen there so much. As I had no other place to go for my social life, these questions distressed me greatly. However, there was no way to sidetrack them.

In one particular aspect, the year 1909 was an *annus mirabilis* at Harvard. I was one of five infant prodigies enrolled as students. The others were W. J. Sidis, A. A. Berle, and Cedric Wing Houghton. Roger Sessions, the musician, entered Harvard the next year at fourteen. I was nearly fifteen at the time, and a first-year graduate student. W. J. Sidis entered college as a freshman. He was the son of a psychiatrist, Boris Sidis, who, together with his wife had opened a private mental hospital in Portsmouth, New Hampshire. Like my father, Boris Sidis was a Russian Jew, and also like my father he had strongly developed opinions concerning the education of children.

Young Sidis, who was then eleven, was obviously a brilliant and interesting child. His interest was primarily in mathematics. I well remember the day at the Harvard Mathematics Club in which G. C. Evans, now the retired head of the department of mathematics of the University of California and Sidis's lifelong friend, sponsored the boy in a talk on the four-dimensional regular figures. The talk would have done credit to a first- or second-year graduate student of any age, although all the material it contained was known elsewhere and was available in the literature. The theme had been made familiar to me by E. Q. Adams, a companion of my Tufts days. I am convinced that Sidis had no access to existing sources, and that the talk represented the triumph of the unaided efforts of a very brilliant child.

There was likewise no question that Sidis was a child who was considerably behind the majority of children of his age in social development and social adaptability. I was certainly no model of the social graces; but it was clear to me that no other child of his age would have gone down Brattle Street wildly swinging a pigskin bag, without either order or cleanliness. He was an infant with a full share of the infractuosities of a grown-up Dr. Johnson.

In his childhood, Sidis had received more than his share of publicity. The papers had a field day when, after one or two years of limited success at Harvard, Sidis received a job at the new Rice Institute in Houston, Texas, under the sponsorship of his friend Evans. He failed to show the maturity and tact needed to make good at this impossible task. Later on, when he carried a banner in some radical procession and was locked up for it, the papers were even more delighted.

Sidis broke down after this episode. He developed a resentment against his family so bitter that he would not even attend the funeral of his father, and a resentment against all mathematics, science, and learning. Indeed, he developed a hatred for anything that might put him in a position of responsibility and give him the need to make decisions.

I saw him many years afterward, when he used to haunt the halls of the Massachusetts Institute of Technology. His intellectual career was behind him. He asked for nothing more than a job at which he might earn his bread and butter as a routine computer, and the chance to indulge his simple amusement of collecting streetcar transfers from all over the world. He avoided publicity as he would the plague.

By this time, the Second World War was approaching, and there was much computing work to be done at the Massachusetts Institute of Technology. It was not difficult to find Sidis a job, although it was always hard to avoid giving him a better and more responsible job than he was willing to undertake. The reports on his work never varied. Within the limitations that he imposed upon himself, he was an exceptionally rapid and competent computer. He had even managed to acquire a certain minimum degree of neatness in his personal appearance, and was a quiet, inoffensive worker to have around. He had a limited security in his work with us; we all knew his story and respected his privacy.

I have no doubt that even at the time I knew him at Harvard, competent psychoanalytic help of the sort that is readily available today could have saved young Sidis for a more useful and a happier career. I am equally sure that his father, precisely because he was a psychiatrist, and was busy reading the fine print of the psychological map, was unable to see the inscription written on it in the largest characters, stretching from one corner to the other. It was perfectly clear that the later collapse of Sidis was in large measure his father's making.

Without condoning the follies of the older Sidis, it is at least necessary to understand them. Picture a Jew, fresh from the persecutions of Russia, and a denizen of a land that has not yet fully made up its mind to want him. Picture his successes, exceeding what he could have hoped for as a child but still falling short of his wishes. Picture the brilliant child, destined for a still greater success, beyond the possibilities open to the parents. Picture the Jewish tradition of Talmudic learning which, from the times of Mendelssohn, has been transferred to the secular learning open to the whole world, and picture the ambition of the orthodox Jewish family to have among their sons at least one great rabbi—and the universal sacrifice which such a family will make to achieve this end.

I am loath to add my name to those who join in an excessive stream of condemnation of Boris Sidis. I have in my possession a letter from a writer who, after spending a day looking over the published accounts of the case, was sure that the father was guilty of a capital crime, and that this crime was the result of the attitude of the scientist who is so devoted to science that he is willing to commit an act of spiritual vivisection on his own child. I think such judgments are premature and are wanting in the sympathy and compassion that mark the really great writer.

I find it appropriate to discuss Sidis in some detail as his life formed the object of a cruel and quite uncalled-for article in *The New Yorker*. Some years ago, when Sidis was leading a broken but independent life around the Massachusetts Institute of Technology, an enterprising journalist got hold of his saga. I believe he won Sidis's confidence. Sidis, who through his later life was a defeated—and honorably defeated—combatant in the battle for existence, was pilloried like a side-show freak for fools to gape at.

He had ceased to be news for nearly a quarter of a century. If any man had done wrong, it was his father, who was long dead, and the article could right no wrong done the son. The question of the infant prodigy was not a live issue, even in the public press, and had not been for some time, until *The New Yorker* made it so. In view of all this, I do not see how the author of the article and the editors of the magazine can support their conduct by the claim that the actions of people in the public eye are the object of fair newspaper comment.

I suspect certain members of *The New Yorker* staff of muddled thinking. In many literary circles, anti-intellectualism is the order of the day. There are sensitive souls who blame the evils of the times on modern science and who welcome the chance to castigate its sins. Furthermore, the very existence of an infant prodigy is taken as an affront by some. What, then, could constitute a better spiritual carminative than an article digging up the old Sidis affair, at the same time casting dishonor on the prodigy and showing up the iniquity of the scientist-prodigy-maker? The gentlemen who were responsible for this article overlooked the fact that W. J. Sidis was alive and could be hurt very deeply.

Sidis sued *The New Yorker* for damages. It is not my object to criticize the courts, and I do not know enough law to describe the case fairly. I believe, however, that the main issue was the fact that to obtain damages in a libel suit where the specific allegations are undisputed and where they concern only the tone of ridicule of the article, it is necessary to prove specific damage which will hamper the injured party in the carrying out of his professional work. Now Sidis had no profession, and the proof of such damages was impossible. He was only a day worker, and it is a fact that no criticism of this sort would have closed his employment or led to a decrease in his wages. This was not the sort of libel case in which personal agony was a legal issue. Thus *The New Yorker* won its case.

When Sidis died some years later, I remember how shocked we were. We tried to obtain from the hospital some indication of the disease that had killed him. We were not kinsmen, and the hospital authorities were quite properly reticent. I do not know to this day what the cause of his death was.

The issue was raised again by an article published in *This Week Magazine* of the Boston Sunday *Herald* in March, 1952. It was entitled "You Can Make Your Child a Genius," and was based on an interview with the mother of William James Sidis. As a piece of reporting, it is an ordinary journalistic hack job, neither better nor worse than a thousand others that appear in the Sunday supplements and the slicks. As a human document, it scarcely merits consideration.

Sidis's failure was in large measure the failure of his parents. But it is one thing to have compassion for an honest human weakness, and it is another thing to brandish before the public the wreck of a human adventure as if it had been a success. So you can make your child a genius, can you? Yes, as you can make a blank canvas into a painting by Leonardo or a ream of clean paper into a play by Shakespeare. My father could give me only what my father had: his sincerity, his brilliance, his learning, and his passion. These qualities are not to be picked up on every street corner.

Galatea needs a Pygmalion. What does the sculptor do except remove the surplus marble from the block, and make the figure come to life with his own brain and out of his own love? And yet, if the stone be spaulded and flawed, the statue will crumble under the mallet and chisel of the artist. Let those who choose to carve a human soul to their own measure be sure that they have a worthy image after which to carve it, and let them know that the power of molding an emerging intellect is a power of death as well as a power of life. A strong drug is a strong poison. The physician who ventures to use it must first be sure he knows the dosage.

The striking thing to many about the group of precocious children who were studying at Harvard in 1909–10 is that we were not an isolated group at all: in some ways we were alike and in some ways different. At least three came from homes of very ambitious fathers, but the fathers were not alike in the slightest degree, nor would their ambition take the same form. My own father was primarily a scholar, and his ambition was for me to excel in scholarship. He took his duty in this matter very seriously, and spent a large if not excessive amount of time in my education. Berle's father wanted his son to be a successful lawyer and a statesman. He took a large share in Berle's early education, but I do not believe that he continued it over the period in which Berle was at Harvard. Sidis's father was a psychologist and a psychiatrist by profession. I have said that he had wished his son to excel in scholarship, although I cannot find in what I remember of young Sidis the same degree of continued participation in his son's education that my father had exhibited in mine. I have no doubt that in Sidis's earlier childhood he had received intense paternal discipline. But during the period in which I first met him when he was about eleven years old, he was left alone in a roominghouse for the greater part of the year in a Cambridge in which he had few personal friends and even fewer intimates.

I know nothing about the relations of either Houghton or Sessions with their families. I presume that part of the reason is that there was less to know and that the families did not participate as intimately and overwhelmingly in their education as did the families of the other three of us. I believe that these boys were left more to their own resources and that consequently they were not subject to the same pressure to which we were subject.

I remember the fourteen-year-old Berle as he first visited me, point-device, carrying his little kid gloves and presenting a formal visiting card. This was a phenomenon new to me, for my academic precocity had prevented neither me nor my parents from being quite aware that after all I was a boy not yet fifteen. Yet I was a physically precocious child, well past the beginning of puberty, while this youngster, although practically of the same age, was in every bodily respect at least five years my junior. To find this poise and sense of social protocol in a half-grown boy was a shock to my sensibilities.

Sidis had his collection of streetcar transfers to amuse him, and Berle had a fad almost as individual. He was interested in the various underground passages of Boston, such as the subways and the sewers and various forgotten bolt holes; and in particular he introduced us to that romantic passageway dating from early colonial times, which still passed in those days under the site of the old Province House. The bricks dated back more than two and a half centuries, and we both showed our essential boyishness by joining in plans for a literary hoax by which we should discover a Shakespearean document buried behind the wall.

I have had no contact with Berle since his graduation from college. He became one of that group of young lawyers and statesmen sponsored by Felix Frankfurter, a group that has been a fertile source of talent. Berle's rise has been quick and sure and in no way surprising, for his ambition was matched by his talents. *The New Yorker* handled him a bit roughly in a profile. But I cannot feel any such indignation at the rough profilement of Berle as I did in the case of Sidis. Berle has been a public figure and he has held a great deal of power. So long as certain canons of journalistic decency were adhered to (and I could not say that they were violated by *The New Yorker*), his acts and personality were of legitimate public interest and the subject of fair comment. Sidis was out of public life and it was an act of the utmost cruelty to summon him back.

We five boys, varying in ages between eleven and fifteen, would not naturally have sought one another's companionship if it had not been for the special circumstances in which we were placed. I have said that Berle and I did not find our first contact particularly stimulating to either of us, and that after our formal introduction to one another we found very little to talk over. Later on we used to bowl together in the bowling alley in the basement of the gymnasium, and once or twice we walked to Boston together. Berle told me something of his interest in the underground passages of Boston, and as I have said, we planned to participate in a joint literary hoax. But our acquaintanceship did not last for it had no roots.

Sidis was too young to be a companion for me, and much too eccentric, although we were in one class together in postulate theory and I respected the work that he did. Houghton was a very good friend of mine and I got to know him the best of the lot. I used to visit from time to time in his rooms in Divinity Hall, and he impressed me as a thoroughly agreeable person. He appeared to have a promising future, which was tragically cut off by a premature death caused by appendicitis at about the time he expected to graduate.

As to Sessions, I met him only once or twice, and the great divergence in our interests prevented us from developing a common vocabulary.

Thus there was little acquaintance and attraction between us as a group. I tried at one time to unite the five of us into a sort of prodigy club, but the attempt was ridiculous for we did not possess a sufficient element of coherence to make a joint social life desirable. We associated with students older than ourselves for intellectual purposes, and with the brighter boys of our own age but of normal advancement in school for the sake of our needs as children and adolescents. In all of our cases, our social relations were better taken care of elsewhere than by a close social contact with those of our own kind. We were not cut from the same piece of cloth, and in general there was nothing except an early development of intelligence that characterized us as a group. And this was no more a basis for social unity than the wearing of glasses or the possession of false teeth. Louise Baker, in her witty book *Out on a Limb,* shows that there is no necessary bond of companionship between one one-legged little girl and another one-legged little girl, and my experience leads me to believe that the sharing of a precocious school career is no sounder a basis for one's companionship than the sharing of a mutilation.

By the end of my first term at Harvard, it had already become more than doubtful whether a career awaited me in biology. As usual, the decision was made by my father. He decided that such success as I had made as an undergraduate at Tufts in philosophy indicated the true bent of my career. I was to become a philosopher and to apply for a scholarship at the Sage School of Philosophy at Cornell University, where my father's old friend of Missouri days, Professor Frank Thilly, filled the chair of ethics. I can understand that with our limited family means and the needs of the other children coming to the front, it was not possible to allow me the chance to make serious errors, but this deprivation of the right to judge for myself and to stand the consequences of my own decision stood me in ill stead for many years to come. It delayed my social and moral maturity, and represents a handicap I have only partly discarded in middle age.

I was, however, not reluctant to leave Harvard. I had felt myself to be a misfit there from the first. Harvard impressed me as being overwhelmingly right-thinking. In such an atmosphere, a prodigy is likely to be regarded as insolence toward the gods. My father's publicly announced attitude toward my education had aroused hostility among his colleagues which made my lot no easier.

I had hoped to find a free intellectual life among my fellow students. Some, indeed, I found who were willing to discuss intellectual matters and to fight dialectically for their convictions. But in the Harvard order of things, a gentlemanly indifference, a studious coldness, an intellectual imperturbability joined with the graces of society made the ideal Harvard man. Thirty years later, I have been shocked rather than surprised by the dry emotional and intellectual sterility into which some of these men have settled.

At the close of the academic year, my father made one decision for which I shall eternally thank him. He rented Tamarack Cottage in the township of Sandwich, New Hampshire, as our home for that summer. Sandwich has remained my summer home for much of the time until the present day, and has a special place in my heart for the loveliness of its scenery, for the walks and climbs afforded by its mountains, and for the sober dignity, reserve, and friendliness of its country people. Some of my walks were short trips to Tamworth or to Sandwich Center, on which I stopped at the backdoors of my neighbors for a chat or a glass of milk or water; some were climbs which my father took with my older sister and me over the trails of Whiteface and Passaconaway. There was one long trip of nearly a week during which my father, my sister, Harold King, and I packed a pup-tent and our duffle bags into the Passaconaway valley, up the logging railways to the remains of Camp Six in the Livermore Wilderness, down by the lumber village of Livermore, to the Notch Road and up the Crawford Trail to the top of Mt. Washington. From there we came down over the heights of Boot's Spur to Tuckerman's Ravine and to the Pinkham Notch, Jackson, Intervale, and Tamworth. We afterward heard that we had passed the home of William James on the day of his death.

Nowadays I spend all of my normal year in New England, my winters in Belmont, Massachusetts, and my summers in Sandwich, New Hampshire, when I am not globe-trotting. Yet, although I spend the greater part of my time in the city, it is my country home to which I feel I most deeply belong. The New Englander is reticent whether he comes from the country or the city; but the reticence of a countryman is most humble and proud. The New Hampshire farmer has a sense of living with his ancestors which he derives from tilling the same land, dwelling in the same house, and often from using the same tools. Yet this sense of historical continuity is too personal and too private to flaunt in the face of the visitor. For the city New Englander, the family is often an asset to the individual; but for the countryman, the individual is but a passing phase in the continuity of the family. If the countryman is reticent it is because he thinks that, important as his affairs may be to himself, it is your own affairs which are most important to you, and he waits for some sign that he is welcome before he intrudes upon you. He waits to look you over, but he also waits to give you a chance to look him over. And even in the meantime, he regards you as a whole individual rather than such an abstract half-man as an employer or a customer. He will not start business with you before he has given you five or ten minutes' gossip, as his right and your own, as two human beings who have more important relationships with one another than that of buying and selling. He will accept a gift but not a tip, a cigar but not a dollar bill. In short, whether you can love him or not, and very often you can, you can and must respect him because he respects himself.

We had many friends in the mountains. One eccentric and interesting neighbor was Professor Hyslop, famous for his work on psychic research. We spent many an interesting hour before the fire in his cottage, a structure which had originally been built for a henhouse, and listened to his weird tales of ghouls and ghosts and mystical noises and mediums. His house belonged to a Mr. Hoag of Philadelphia whose family still inhabits the region into the third generation. The Hoag boys, young Hyslop, and I belonged to a very informal baseball team which also comprised the son of Professor Dugald Jackson of the Massachusetts Institute of Technology, two sons of the late President Grover Cleveland, and one or two of the Finlay boys from New York. I do not remember the names of the other players. We used to practice on a none-too-horizontal field near the Finlay home, and the practice always meant for me a five-mile walk there and back. We played no more than two games, both of which we lost ignominiously. I was a substitute on the team. I think nothing can express better my exact status as an athlete.

« X I »

DISINHERITED

Cornell, 1910–1911

I HAD WON a scholarship at Cornell. Father was to accompany me to Ithaca, and at the end of the summer we had to decide how to get there. This was still in the days before the interurban trolley had been superseded by the bus and the coach. Father and I decided on a romantic trolley jaunt to central New York and Ithaca. There we called on Professor Thilly and made plans for the ensuing academic year. It was decided that I was to have the free run of the Thilly house, and to confide my youthful troubles and perplexities to Professor Thilly and his wife.

My father and Professor Thilly had a long evening talk together about the old days at the University of Missouri and about many other matters. In the course of the rather miscellaneous discussion, Thilly casually mentioned to Father that he remembered an allusion many years ago to a much earlier philosopher in the family, Maimonides. Father admitted hearing of rumors, perhaps not authentic but depending on old documents that my grandfather had lost, to the effect that we were descended from Maimonides.

I had not previously heard of the tradition, nor even of the name of Maimonides. Naturally I did not delay long before I had recourse to the encyclopedia. I found there that Maimonides, or Rabbi Moses ben Maimon, known as Rambam according to the conventional Jewish use of initials, was a native of Cordova domiciled in Cairo, and the body physician to the vizier of the Sultan Saladin. I learned that he was the head of the Jewish community of Egypt and a great Aristotelian and that his most famous book is known as *Moreh Nebukim*, or the *Guide of the Perplexed*.

I was naturally interested to have such an important figure on which to hang our family pride, but the implications of the legend came to me with a profound shock. For the first time, I knew that I was Jewish, at least on my father's side. You may ask how it was possible for an intelligent boy like me to have any doubts about this when my grandmother Wiener as far back as I could remember had received a newspaper printed in what I knew to be Hebrew characters. I can only answer that the world is complex, with ramifications not very understandable to an adolescent, and that it still seemed possible to me that there might be non-Jewish people in eastern Europe who used the Jewish characters. Furthermore, my cousin Olga had once told me that we were Jews; but my mother had contradicted this at a time when I had not yet learned to question the word of my parents.

At that time the social disadvantage of belonging to the Jewish group was considerably greater than it is now and there was definitely something to be said for allowing children to grow up through their early lives without consciousness of the social stigma of belonging to an unfavored group. I do not say categorically that this was the right thing to do; I merely say that it was a defensible thing to do and could be motivated

—in fact, it was actually motivated—by a desire for the protection of the children. The moral responsibility of a policy like this is great. It is done nobly or it is done basely.

To put the best possible light on this course of action, it would be necessary that the children brought up in ignorance of the fact that they historically originated from the Jewish group be also brought up with an attitude of understanding. They should be made to see that the disadvantage on the part of others of belonging to such an unfavored minority group is an unmerited burden which they should at least abstain from intensifying. Such an attitude should be carried throughout the whole of life and should be directed equally against the unjust stigmatization of Jews, of Irishmen, of recent immigrants, and of Negroes, et cetera. Of course, the best thing and indeed the only one that would be thoroughly justifiable from a moral point of view would be to excite in the child a resistance, if not hostility, to all forms of belittling prejudice, no matter what the objects might be. However, short of this, every word by which the child's prejudice may be excited or intensified is a blow against his moral integrity, and ultimately a blow against his confidence and belief in himself when he should come to know, as he inevitably must, the truth of his own origin. The burden of the consciousness of Original Sin is hard enough to bear in any form; but a particularly insidious form of it is the knowledge that one belongs to a group that he has been taught to depreciate and to despise.

The responsibility for keeping the fact of my Jewishness secret was largely my mother's. My father was involved in all this only secondarily and by implication. I believe that he had originally intended not to burden us by the consciousness of belonging to an undervalued group, while at the same time he wished to preserve intact our respect for that group and our potential self-respect. He had written a number of articles on Jewish themes as well as a *History of Yiddish Literature*. He was also the first person to bring the name of Moritz Rosenfeld to the attention of the non-Jewish public. Father had been engaged in various negotiations with the Jewish Publication Society and with other similar Jewish organizations, and I gather these had involved considerable friction. Later I found that Father always claimed that the friction was the result of an arrogant insistence on the part of the Jewish organizations that a Jew was a Jew before he was a man, and that he owed inalienable allegiance to his own group before humanity itself. My father was always an individual, and was the last man in the world to stand pressure of this kind.

My mother's attitude toward the Jews and all unpopular groups was different. Scarcely a day went by in which we did not hear some remark about the gluttony of the Jews or the bigotry of the Irish or the laziness of the Negroes. It is easy enough to understand how these sops to the prevailing narrow-mindedness of that epoch were thrown out by one who had experienced the disadvantages of belonging to an unfavored group; but though one can understand the motives leading to this conformist spurning of one's own origins, and can even forgive it in the sense in which the religious man hopes that his sin will be forgiven him, one cannot help regretting it and being ashamed of it. He who asks for equity must do equity, and it is not good for the children of a Jewish family, whether they know they are Jewish or not, to hear another Jewish family spoken of contemptuously for making the same efforts to pass over that their own parents have been making.

The maintenance of a family silence such as that which my parents maintained, even if it might be considered advisable, is much more difficult than it appears at first glance. If there is agreement that such a silence shall be maintained, what can one partner do when the other makes disparaging remarks about the race before the children? He can either terminate the secret, or he can be a silent and unwilling observer of a course of conduct that can lead to nothing but delayed emotional catastrophe on the part of the child. The vital danger of even the whitest of lies is that if it is to be maintained it must lead to a whole policy of disingenuousness of which the end is not to be seen. The wounds inflicted by the truth are likely to be clean cuts which heal easily, but the bludgeoned wounds of a lie draw and fester.

In offering the maximum apology which I can for the course of conduct adopted by my parents, I do not wish to justify it as a whole or to condemn it. I do mean to affirm that it had serious consequences for me. I was led very quickly into a rebellion against my parents and to an acceptance of their disfavor. Who was I, simply because I was the son of my mother and father, to take advantage of a license to pass myself off as a Gentile, which was not granted to other people whom I knew? If being a Jew was something to invite a shrug of the shoulders and a contemptuous sniff, then I must either despise myself, or despise the attitude by which I was invited to weigh myself with one balance and the rest of the world with another. My protection may have been well intentioned, but it was a protection that I could not accept if I were to keep my integrity.

If the maintenance of my identity as a Jew had not been forced on me as an act of integrity, and if the fact that I was of Jewish origin had been known to me, but surrounded by no family-imposed aura of emotional conflict, I could and would have accepted it as a normal fact of my existence, of no exceptional importance either to myself or to anyone else. Probably some conflict would have been excited by the actual external anti-Semitism which I found to belong to the times, and which would sooner or later have hit me one way or another. Nevertheless, unless there had been an ambiguity in the family attitude, this would not have hit me where it really hurt, in the matter of my own internal spiritual security. Thus the effect of an injudicious attempt to conceal from me my factual Jewish origin, combined with the wounds which I suffered from Jewish anti-Semitism within the family, contributed to make the Jewish issue more rather than less important in my life.

I say these things with the very clear and explicit intent to help prevent those who may be tempted to repeat this mistake of starting the child into the hurly-burly of life with this unnecessary sense of frustration and damnation.

Thus, when I became aware of my Jewish origin, I was shocked. Later, when I looked up my mother's maiden name and found that Kahn is merely a variant of Cohen, I was doubly shocked. I was not able to defend myself by the divided personality which allowed my mother's family to weigh strangers and their own kinsmen with different weights. As I reasoned it out to myself, I was a Jew, and if the Jews were marked by those characteristics which my mother found so hateful, why I must have those characteristics myself, and share them with all those dear to me. I looked in the mirror and there was no mistake: the bulging myopic eyes, the slightly everted nostrils, the dark, wavy hair, the thick lips. They were all there, the marks of the Armenoid type. I looked at my sister's photograph, and although she was a pretty girl in my eyes, she most certainly looked like a pretty Jewish girl. She had features not unlike those of a Jewish boy who happened to be my fellow lodger in my Cornell boardinghouse. He was a member of a recent immigrant family and appeared very foreign against the background of his Anglo-Saxon classmates. My snobbishness prevented me from accepting him fully as a friend, and the meaning of this was clear to me: I could not accept myself as a person of any value.

In this emotional and intellectual dilemma, I did what most youngsters do—I accepted the worst of both sides. It humiliates me to think of it even at the present day, but I alternated between a phase of cowardly self-abasement and a phase of cowardly assertion, in which I was even more anti-Semitic than my mother. Add this to the problems of an undeveloped and socially inept boy, spending his first long visit away from home, to the release from the immediate educational pressure of my father before I had developed independent working habits, and you have the material of which misery is made.

For I was miserable. I had no proper idea of personal cleanliness and personal neatness, and I myself never knew when I was to blurt out some unpardonable rudeness or *double entendre*. I was ill at ease with my associates in the middle twenties, and there were no youngsters of my own age to replace them. The habits of vegetarianism inculcated in me by my father had increased the difficulty of a social life away from home, and together with other people. Yet I was greatly under his spell, and because of my upbringing, even the remote threat of his powers made me loath to abandon these habits as my sisters were later to do.

My studies at home had always been under the close inspection of my father. This made it hard for me to develop good independent habits of study. I know that Father claimed to have always been in favor of my intellectual independence, and to have wished me to stand firmly on my own feet; yet whatever he purported to wish, the pattern of our life together worked in exactly the opposite direction. I had grown to depend on his support, and even on the support of his severity. To pass from this shelter to the full responsibility of a man among men was too much for me.

I took a rather divers series of courses my year at Cornell. I read Plato's *Republic* in Greek with Hammond and found that I had not lost too much of my Greek fluency in my Harvard year. I also attended the psychological laboratory and took a course with Albee on the English classical philosophers of the seventeenth and eighteenth centuries. Albee's course was dry but instructive, and I believe that there is an element still in my literary style which I owe to the rapid perusal of a large amount of seventeenth-century material.

I tried to take a mathematical course under Hutchinson on the theory of functions of a complex variable, but I found that it was beyond me. Part of the difficulty came from my own immaturity, but another part—to my way of thinking—from the fact that the course did not cover an adequate approach to the real logical difficulties of the subject. It was only later at Cambridge, when I found these difficulties faced boldly by Hardy instead of left to the student with an attitude of "Proceed and faith will come," that I began to find myself at home in function theory.

I did not fall down so badly in the Plato course, for this was but the continuation of my father's teaching under another preceptor. In my metaphysical and ethical courses I suffered from a new and vague adolescent religiosity (which did not last very long), and it needed a sharp logic to keep me from diffuse sentimentality.

I had to write essays for Albee on the seventeenth- and eighteenth-century philosophers. I was cramped by a boyish style, together with a physical awkwardness with the pen. My essays wound up in compact knots of words, so at variance with the norms of the English language that I was more than once asked whether my first language had not been German.

Cornell University had a philosophical periodical of its own, and one of the duties of the Sage Fellows was to write brief abstracts of articles in other philosophical journals to be included in a special section devoted to the purpose. The original languages were English, French, and German; and the exercise of translation familiarized us both with the philosophical vocabulary of these languages and with the ideas current in the world. I cannot vouch for the quality of our translations, but I have a very profound impression of the intellectual value of the work to ourselves.

There were some reliefs in this black year of my life. Although I could not fully share the companionship of my fellow students, the picnics up a neighboring creek and the winter sleigh rides which took place there after the fall of the snow were very pleasant indeed. There were one or two undergraduates in my boardinghouse with whom I had good times and bull sessions, and they used to play childish pranks on me and on one another. The scenery of the campus was gorgeous and later, when spring came around, the plantations of flowering quince were beautiful beyond anything I had seen on the Tufts College campus or elsewhere. There were sailboat rides on Cayuga Lake and tramps to neighboring waterfalls, where we swam and bathed under the plunging masses of water.

I have carried down to the present day the friendship of more than one of my fellow students. Christian Ruckmich, a gaunt, Lincoln-like figure, was my partner in many of my walks and in the psychological laboratory. I have heard from him from Abyssinia within the last few years. He has been engaged there in reforming the educational system of the country, and his boy has taken up aviation.

There was also Tsanoff, the Bulgarian, whom I have seen within the last year or two at the Rice Institute, still teaching philosophy. There was a delightful couple by the name of Schaub with whom I often lunched. Schaub taught a course on comparative religion, and his discussion of the Old Testament fitted in very well with the philological interest which I had acquired from my father, from Professor Wade at Tufts, and from browsing around our library.

As the year went on, it began to appear that I had not earned a renewal of the fellowship which had taken me to Ithaca, or at any rate if I were to receive it again, it would be by special grace. I felt oppressed, not only by my indifferent success in my courses, but by that sense of adolescent guilt that accompanies the internal sexual development of almost every normal young person. My sense of guilt led me to avoid the Thillys, and this alienation ended in a quarrel between my father and Professor Thilly. It was almost impossible to make my father believe that one of his family could be at fault. It was even more impossible for me to stand up to the withering stream of invective which a discussion was bound to bring down upon my head.

Before the end of the year there was other news from home. I had a new brother, a sickly child, who scarcely lived out the year. With the bad news from Cornell, my father snatched me out of the Sage School and forced me to transfer to the department of philosophy of the Harvard Graduate School. I know that the responsibility of my family made it difficult for my father to back a doubtful bet, but I nevertheless wish that as a young man I had received the opportunity to redeem an error in the place in which I had made it. The result of Father's policy of transferring me was to increase my lack of self-confidence, which was already great enough. My blunders became a menacing sequence of dead years never to be undone. Meanwhile I did not have the opportunity to learn the arts and techniques of independence, and the future was for me a turbid and depressing pool.

I had time after I came home to take stock of my moral situation. My achievement of independence during the year at Cornell had been incalculably retarded by the confused mass of feelings of resentment, despair, and rejection which had followed early in the year upon the discovery of my Jewishness.

Some of my friends have asked me to render more specific the discussion of the shock which I experienced and the subsequent adjustment which it was necessary for me to go through in order to be reasonably at peace with myself. Manifestly to be at once a Jew and to have had inculcated in me by certain of my mentors hostile or depreciating attitudes to the Jews was a morally impossible position. It might have led me into a continued Jewish anti-Semitism or, on the other hand, to a flight into Abraham's bosom.

In fact, neither of these escapes was possible for me. I had received too strong a lesson in intellectual and moral integrity from my father to be willing to accept one brand of justice for myself and my near kinsfolk and another brand of justice for the outer world. I had heard enough harsh remarks at home concerning other university families of Jewish origin who had tried to escape from Judaism to realize that there were those close to me who weighed the Wiener family with one scale and the rest of the world with a very different one. Quite crassly, even if I myself and some of my immediate family had been willing to deny my Jewish origin, this denial would not pass current one foot beyond our doorstep.

In short, I had neither the possibility nor the wish to live a lie. Any anti-Semitism on my part must be self-hatred, nothing less. A man who hates himself has an enemy whom he can never escape. This way there lay only discouragement, disillusionment, and in the end madness.

Yet it was equally impossible for me to come into the fold of Judaism. I had never been there, and in my entire earlier education I had seen the Jewish community only from the outside and had the very vaguest idea of its rites and customs, its permissions and obligations. The break with orthodox Judaism had indeed begun in my grandfather's time; in pursuit of his desire to Germanize the eastern Jews and to replace Yiddish by High German, my father had been sent in part at least to a Lutheran school. Thus a return to Judaism on my own part would not be a true return, but a fresh conversion and conviction. For better or worse, I do not regard conversions of any sort with very favorable eyes, nor did my father. There is something against the grain in the attitude of abnegation and of denial of personal judgment in the wholesale acceptance of any creed, whether in religion, in science, or in politics. The attitude of the scholar is to reserve the right to change his opinion at any time on the basis of evidence produced, and I was born and bred to the scholar's trade.

This training of mine went very deep into my nature. I have never had the impulse to gregariousness in my thinking and feeling despite all my very deep respect for man as man, whether he be a scholar or not. It was emotionally impossible for me to hide myself in the great majority as a fugitive from Judaism; but it was equally impossible for me to hide myself and be consoled in a restrictedly Jewish community. I could not believe in the old-line New Englanders as a Chosen People: but not even the vast weight of the Jewish tradition could persuade me to believe in Israel as a Chosen People. The one thing that I had known about my father's relation to Judaism was that he had been an assimilationist, rather than Zionist, and that he had had many arguments with Zangwill and others on this issue. This was a position which I approved not only because he was my father but also because I thought he had seized the stick by the right end.

Thus I was powerfully moved by the discovery of my Jewishness, but I could see no way out in anti-Semitism or in ultra-Judaism. What, then, could I do?

I cannot tell when I arrived at an answer to my problems, for the solution occurred step by step and was not reasonably complete until after my marriage. Yet one thing became clear very early: that anti-Jewish prejudice was not alone in the world but stood among the many forms in which a group in power sought, whether consciously or unconsciously, to keep the good things of the world for itself and to push down those other people who desired the same good things. I had read

DISINHERITED

enough of Kipling to know the English imperialist attitude, and I already had enough Hindu friends to realize how bitterly this attitude was resented. My Chinese friends spoke with me very frankly concerning the aggressions of the Western nations in China, and I had only to use my eyes and ears to know something of the situation of the Negro in this country, particularly if he aspires to be something more than a farm hand or an unskilled worker. I was quite adequately informed concerning the mutual bitterness between the old Bostonian and the rising Irish group which demanded its own share of power in the community and took a very liberal view of what that share should be when other immigrants and minor groups came into question.

The net result was that I could only feel at peace with myself if I hated anti-Jewish prejudice as prejudice without having the first emphasis on the fact that it was directed against the group to which I belonged. I felt anything less than this as a demand for special privilege by myself and by those about me. But in resisting prejudice against the Oriental, prejudice against the Catholic, prejudice against the immigrant, prejudice against the Negro, I felt that I had a sound basis on which to resist prejudice against the Jew as well. For a long time I had been interested in my fellow students from the Orient and from other foreign countries, and I now saw their problems as parallel to my own and, in many cases, far deeper and more difficult.

Moreover, when I heard of our reputed descent from Maimonides, I realized that even deeper than our simple Jewishness, in a sense the Orient was part of our own family tradition. Who was I, a man whose proudest ancestor had led a life in a Moslem community, to identify myself exclusively with the West against the East? Thus I came to study and to observe parallelisms between the intellectual development of the Jews, especially in that interesting period of transition which began with Moses Mendelssohn and led to the integration of Jewish learning with European learning in general, with similar phenomena taking place before my eyes among the non-European men of learning. This came to an even sharper focus later on when I spent part of a year assisting Professor Hattori, a Japanese professor at Harvard, in the routine work of his course on Chinese philosophy.

This covers the intimate personal side of my reaction to my discovery that I was of Jewish origin. It may be well, however, to add a few facts concerning anti-Semitic prejudice and its history in those communities in which I have lived since my childhood. It is fairly clear from the history of those Jewish families who came to the United States before the middle of the last century that anti-Semitic prejudice was not a considerable factor in their lives. As a matter of fact, the dominating Protestants in the United States were more than ready to acknowledge the extent to which they had drawn upon the Old Testament in their writings and thought and to see in the Jewish immigrant a reflection of their own traditions. I have been told that even the "Know Nothing Movement" was not particularly anti-Semitic and, further, that some of the leaders of this unsavory episode in our history were drawn from the Jews themselves. Be that as it may, the beginning of the twentieth century saw the blunting of our national resistance to anti-Semitism as it saw the blunting of New England's traditional friendship for the Negro and of many other of the broader attitudes of earlier days. The Gilded Age had already come to an end and had left as its heir the Varnished Age.

« XII »

PROBLEMS AND CONFUSIONS

Summer, 1911

THAT SUMMER we spent at a farmhouse not far from Bridgewater, New Hampshire. There was only one small mountain in the immediate neighborhood, and it was too rough and trailless for my father to permit me to climb it. I tramped the roads of the neighboring countryside in search of summer camps where I might earn a little money as a teacher, and find a little companionship in the bargain, but my services were not in demand. I pitched hay in a local field and fell desperately ill under an allergy to haydust. I read back numbers of *Harper's* and *Scribner's Magazines* and the *Century,* and I longed for the beginning of term to relieve me from the boredom which came from a family living too close together and driven in upon itself.

My father's revolutionary theories of education were confirmed in his eyes by the success which, with all my shortcomings, I had already found in intellectual work. It soon became clear that my sisters, although very clever girls by any ordinary standard, were not responding to my father's training as I had. And in part, my father did not expect as much of them. This was laid to their being girls, unable to stand up to the severe discipline to which I had been subjected.

Our family portioned out the fates of the family members in advance. The expectation that my sister Constance was to be the artistic one made my parents assign music, painting, and literature to be her field. To prevent any contretemps, they were to be strictly eschewed by the rest of us.

Thus Constance, and, in a similar way but later, my sister Bertha too, was removed from the sphere of intellectual competition into which I had been initiated. Occasionally I indulged in a certain envy of their easier fates, and there were times when I would have considered it a privilege to be born a girl and to be faced no longer with a need for the hard work of intellectual effort, and the ultimate requirement of standing alone in a world I felt to be hostile.

The case of my brother Fritz was of course a very different matter from that of my sisters. It was not until I was a graduate student at Harvard that he had reached the age where his education severely impinged upon us. He was destined by my parents for the same career of scholarship as I. This time there was no question of weaker demands on the weaker sex, and my father's educational theories had to be faced in their full significance. My father had reiterated that my success, if indeed I had had any genuine success, was not so much a result of any superior ability on my part as of his training. This opinion he had expressed in print in various articles and interviews.[1] He claimed that I was a most average boy who had been brought to a high level of accomplishment by the merit of his teaching and by that

[1] An article entitled *"New Ideas in Child Training"* by H. Addington Bruce, published in *The American Magazine* in July, 1911, quoted my father directly:
    I am convinced that it is the training to which we must attribute the results secured with them. It is nonsense to say, as some people do, that Norbert and Constance and Bertha are unusually gifted children. They are nothing of the sort. If they know more than other children of their age, it is because they have been trained differently.

PROBLEMS AND CONFUSIONS

merit alone. When this was written down in ineffaceable printer's ink, it had a devastating effect on me. It declared to the public that my failures were my own but my successes were my father's.

Now that my brother had come of school age, there was a second Wiener candidate for fame and distinction, and for the upholding of the judgment of my father. It was inevitable, and it had been made publicly inevitable, that my father should try to duplicate with his younger boy what he had already accomplished with me. It became almost as inevitable that the anticipation of Fritz's success was to be thrown in my face in order to deflate me and to exalt the authority of my father.

I never agreed with Father in his estimate of me as a boy of average abilities, which I always felt had been adopted to curb my self-conceit and cut me down to family size. It was not fair to expect a priori that Fritz could do what I had done. Furthermore, Father did not take into consideration the fact that although I was a nervous and difficult youngster, I had plenty of stamina, and could absorb without utter destruction a punishment far greater than that which the average child could take. Thus when my brother turned out to be a somewhat frail child, endowed with what I believe to be good average ability but without any exceptional powers, the scene was set for trouble.

The bickering about Fritz's education lasted for well over twenty years. I resented as unfair the extra weight which my parents threw into the scales to equalize the balance between my brother and me. I was also very much displeased with the role that I was given at sixteen as my younger brother's mentor and nursemaid, taking him to primary school every morning before beginning a full day's work. I was expected to display to him a companionship rarely to be shown by a lumbering adolescent to a mere child eleven years younger. This age difference was critical. When I was sixteen, he was five; when I was twenty-five, he was still only fourteen.

In defense of my parents' expectations of my relationship with Fritz, it is necessary to remember that the world was changing even during that period before the First World War when I was growing up. When I and my older sister had been young children, not even the relative poverty of the family had kept my mother from having the assistance of at least two servants, one of whom had been a cook and the other generally an excellent nursemaid. The changes of the century had already begun to dry up the stream of immigrant household labor, and wages had risen sharply. Not even our greater prosperity could make up for the new conditions and re-create a class of labor that had almost ceased to exist. Thus the care of the younger child fell on the older one.

Looking back on the matter from my present point of view, I cannot blame my parents for passing on to me a responsibility which they had taken so readily in the care of the elder children, yet the circumstances of my responsibility were not fair to me. My duty to Fritz was a deputized one, completely unaccompanied by authority. Fritz was tiresome under my tutelage, and if I took any measure, no matter how mild, to make him behave, he had only to complain to our parents. Whatever step I might have taken was inexcusable to them. Furthermore, I was a confused, socially inept adolescent, who by any reasonable standard had been overworked for years and who needed every available moment to develop his social contacts and his social poise.

It will be no surprise that my companions, whether boys and girls, whether men and women, were judged more critically by my parents according to whether they accepted or did not accept Fritz than on any other point whatsoever. This, too, was unfair to me. It is too much to ask young people to take as a friend another youngster who always has an infant brother toddling after him, particularly when he has no authority over his brother and the child knows this. Thus there is plenty of explanation, if not excuse, for the fact that I was often harsh if not cruel to my brother. Irony and sarcasm are the weapons for those who have no other weapons; and these I did not spare. The difficult situation grew even more difficult.

To a certain extent, too, I was given the responsibility but not the authority for Fritz's education. Fritz rapidly developed a vocabulary of the intellectual, far beyond his understanding. In the competition within the family, he tried to hold his own by asking questions of a learned sort, with answers he did not fully understand and in which he had no deep interest. I was told that I was to answer these questions in detail, even when they had ceased to have an interest to Fritz and when his mind was wandering elsewhere. When the family went to the theater together, I was supposed to offer a commentary on whatever features of the show had excited my brother's desire to display his intelligence, and I did not have a decent opportunity for the reflections on it which belonged to me for myself and for discussions with my true intellectual contemporaries.

Of course, in all this I am going well beyond the period of my history that I have made the subject of this chapter and I am giving an account of a festering sore which continued to infect our family life. During a considerable part of this period I was living at home, either as a minor or as an adult paying his contributions to the family fund. It may be asked why I did not leave the family to take lodging somewhere for myself, perhaps even in Cambridge. Many times I was on the point of doing so, and many times my parents indicated that if I continued my course of conduct, this would be the inevitable result. However, it was made clear, particularly by my mother, that the separation would be held against me for all eternity, as a sign of my ultimate failure, and would mean the complete and final collapse of family relations.

During the earlier period of my life at home, I was made aware that I was completely dependent on the family bounty, and that such sums as I gained by scholarships were only a partial offset against this. Later, when I had acquired the ability to earn my own living, I had still not acquired any circle of friends outside the family. Thus, while separation from the family might have been desirable, exile from the family was exile into outer darkness.

Those who read further in this book will see that my summers were marked by long mountain excursions for many years before my marriage. Later on, these excursions were replaced to some extent by trips to Europe, often together with my sisters. These gave some alleviation of the pressure of family life, and in particular of my forced custodianship of my brother, and were absolutely essential to my well-being. However, my parents made every attempt to compel me to accept Fritz as a companion on my mountain hikes. This was inequitable, and represented a demand that I could not accept.

This had not been the first time that the rather patriarchal structure of the family had disturbed me. Once, in my youth, my father had planned to join with me in making a museum collection of the fauna and flora of the Old Mill Farm neighborhood and had proposed that we spend a large part of our spare time in maintaining this collection. Once he had spoken of his intention, when Constance and I had grown up, to devote the rest of his life to the conducting of a children's school on his own principles, in which my sister and I were to be teachers. More than once he had talked of returning to the romantic adventure of his youth, and of crossing the continent with us in a covered wagon. All these projects were admirable as indications of his youthful spirit, and would have been most charming suggestions of paternal love and family interest in a household less strictly under parental control. As things were, they represented another turn of the screw.

The summer always found us raising a garden, and I was relegated to the tasks of weeding, thinning asparagus, picking peaches, and the like. These were light tasks, and would have been most agreeable if they had not represented a simple outdoor extension of the regime of my filial servitude. I was clumsy and I was inefficient and I was lazy; and I had to hear these faults ding-donged into me hour after hour as I worked by my father's side in the fields. I was quite as unsatisfactory a farmhand as my father indicated, and I certainly developed a repugnance for work in the fields. This has lasted to the present day, and has hampered me at a time when my diminishing stock of muscular vigor would otherwise make me welcome light garden work as a still admissible form of bodily activity. Nevertheless, so long as my father's mentorship continued to dictate my way of life all winter, it was intolerable that the summer, which I needed badly for recuperation and the formation of new social contacts, should merely be an extension of the winter regime.

Indeed, at a later period after the First World War, when father had sold the Sparks Street house as too large for a family no longer needing growing space, he put the money not only into a smaller and older house on Buckingham Street, but into an apple farm in the town of Groton, Massachusetts. He had hoped that the whole family would co-operate in this work, at least during the apple-picking season, and that the place would furnish in return a cool summer for the married children and the expected grandchildren. The scheme was bound to fail from the beginning. Young people in their early twenties have to consider the pressing necessity of their own social life, and cannot long be denied the opportunity for seeking and meeting their prospective mates.

«XIII»

A PHILOSOPHER DESPITE HIMSELF

Harvard, 1911-1913

I RETURNED to Harvard as a candidate for a Ph.D. in Philosophy in September, 1911, when I was nearly seventeen years old. The period between 1911 and the completion of my doctorate was the period of great names in the Harvard philosophy department, and although William James was dead, Royce, Palmer, Münsterberg, and Santayana were alive and active.

In my first year I took a course with Santayana. I remember very little of its content but a considerable amount of its atmosphere. The feeling of a continuity with an old culture and the feeling that philosophy was an intrinsic part of life, or art, and of the spirit, gave me a great deal of satisfaction; and yet, after the passage of all these years, I cannot put my finger on any definite idea which that course has given me.

Palmer's courses mean as little to me in retrospect. They were reading courses, and as far as I remember, they covered the traditional philosophy of the English school. What I do recollect of Palmer was his grave, sweet-tempered personality,

A PHILOSOPHER DESPITE HIMSELF

a little bowed down by the weight of years, but still eager to encourage the ideas and to allay the natural timidity of the young student.

Ralph Barton Perry was among the chief of those to welcome me on my father's behalf as a student. He and Holt, the psychologist, were two among the five or six authors of a then popular manifesto known as the "new realism." It contained the combined echoes of James's pragmatism with a certain analogy with the work of Bertrand Russell and G. E. Moore in England, and represented a protest against that idealism that dissolves all things into mind and the phenomena of mind. It made a rather plausible case for itself, yet I remember that its chief impression on me was of an intolerable shoddiness and brashness. One of the authors went so far as to try to base his ideas on a mathematical logic in which every other word represented a confusion of misunderstood terms. The literary style of the composition was sophomoric. Nevertheless, I remember Holt as a brilliant and charming personality, as a fluent dialectician within the frame of his seminar, while Perry has remained one of the great and dignified figures of American liberalism.

I had two different sorts of contact with Josiah Royce. One was in his course on mathematical logic. Although I did not regard his own contributions to mathematical logic as of major character, he introduced me to the subject. Royce was a many-sided man, coming at a critical period in the intellectual world when the old religious springs of philosophical thought were drying up and new scientific impulses were bursting into life. His mathematical logic bore the signs that almost always indicate a brilliant man who has come too late into a new field to obtain perfect mastery of it.

This position of facing both the past and the future was also clear in Royce's seminar on scientific method, which I attended for two years, and which gave me some of the most valuable training I have ever had. Royce welcomed into this little group every sort of intellectual who was carrying out a reasonable program of work and who was articulate concerning the methods by which he had come to his own ideas and concerning the philosophical significance thereof.

The group was heterogeneous to say the least. Among us was a Hawaiian expert on volcanos. He has left on my memory only the impression of the two words *pahoehoe* and *aa*, which I understand to be the designation of two types of lava. Among us was also Frederic Adams Woods, the author of *Heredity in Royalty*, a eugenicist of a snobbish cast of mind and a pregenetic point of view. Percy Bridgman, who was even then beginning to be skeptical about the elements contained in experiment and in observation, and who understood the influence on physics of James's pragmatism, was definitely veering toward the operational position which he later assumed. The first head of Boston's Psychopathic Hospital, Southard, spoke interestingly on the problems of psychiatric method. There was also Professor Lawrence J. Henderson, the physiologist, who combined some really brilliant ideas about the fitness of the environment with what seemed to me a distressing inability to place them in any philosophical structure, and whose pomposity of manner was not diminished by the article of faith that led him to place a great business administrator in the scheme of things about halfway between a pure scholar like himself and the Creator. Incidentally, I found that those who undervalue their own profession of learning are rarely those who rise to the greatest heights in it.

I believe that it was in this seminar that I first met F. C. Rattray, an Englishman, later to become a Unitarian clergyman and to occupy the pulpit of a church in the English Cambridge. At that time it was Rattray more than any of my official teachers who showed me what good dialectic was, and to what fineness the art of class discussion could be carried.

A PHILOSOPHER DESPITE HIMSELF

I have never seen a man as able to prick a bubble in the froth that often attends such arguments. Nevertheless, I could not help feeling that his devotion to Samuel Butler and his life force, after the fashion of Bernard Shaw, was more a matter of personal emotion adroitly defended by a keen wit than something susceptible to rigorous argument. Rattray and I often used to join forces in the seminars in which we participated, and I am afraid that I became an apt pupil of his and a thorn in the flesh of my mentors.

I also attended Münsterberg's seminars. He was a most puzzling personality. How much of his arrogance was a covert contempt for the America in which he was teaching, and the result of comparison of it with the Germany in which he had failed to find permanent lodgment, we shall never know. His transpontine personality was curiously modeled after that of the German Kaiser, and in my opinion was not a little expressive of the unsureness and brusque assertiveness which went as a false streak through many different social layers of the powerful and able Second Reich. Whatever his private opinion of the America which he adopted and which had adopted him, he had become the master of one of its best rewarded arts: that of personal publicity. His portentous interviews were given a much more intriguing tone by the heavy foreign accent and the slightly foreign phraseology in which they were presented, and Münsterberg became the joy of the reporters.

I learned the mathematical aspect of my philosophy from Professor E. V. Huntington. He was an old friend of my father, and had visited us when we were living in Old Mill Farm in the town of Harvard. I remember that at that period, before I had been graduated from high school, Huntington had tried me out with a little analytical geometry, and had shown me the theory of the nine-point circle.

Huntington was a magnificent teacher and a very kind

man. His exercises in postulate theory were all educational gems. He would take a simple mathematical structure and write a series of postulates for it for which we were to find not only examples satisfying the complete list, but other examples failing to satisfy it in just one place or in several specified places. We were also encouraged to draw up sets of postulates of our own. Both Sidis and I were in this class, and it was here that I first became aware of the boy's real ability and of how great a loss mathematics suffered in his premature breakdown.

Huntington's career has always remained a mystery to me. With his keenness and his inventiveness I should have expected some great mathematical contribution from his pen. Nevertheless, all his pieces of work, no matter how much they may contain of ideas, have remained miniatures and vignettes. I do remember one larger piece of work of his in which he attempted to give a basis to plane and solid geometry; but on the one hand, the work did not go greatly beyond the slightly earlier efforts of Hilbert, and on the other, some of his chief ideas had already found a representation in the work of Whitehead. The valuable and honorable career of Huntington seems to me to bear the lesson that one of the most serious possible lacks in mathematical productivity is the lack of ambition, that Huntington simply set his sights too low.

Let me say a word or two about my amusements during these years. During my long mountain tramp with my father in 1910, I had become acquainted with the excellent work done by the Appalachian Mountain Club in the maintenance of trails in the White Mountains. I joined the club in the fall of 1912, and I got a large part of my exercise in its Saturday walks. A group of us, distributed in age and sex, but all of us devoted trampers, were in the habit of foregathering at one of

A PHILOSOPHER DESPITE HIMSELF

Boston's railroad stations for a suburban train trip and an afternoon's brisk walk in the country.

In 1912 I had obtained my M.A. It did not represent any particular stage in the voyage leading to the Ph.D. but it was convenient to have in case I should meet any obstacle the next year. I had also passed, as I have said, the preliminary examinations on a variety of topics, and they threw me into somewhat closer contacts with my fellow students than I had found already.

Among other things, this was the year of the *Titanic*. It represented a shock to our emotional security which was a fitting introduction for the great shocks to follow. It was perhaps this event rather than the beginning of the First World War two years later that awakened us children of the long peace that had so long protected Europe and America to the fact that we were not the favored darlings of a beneficent universe.

Besides my usual reading from Dumas and Kipling, who were the delights of an adolescent boy, I added to the list of the books that I particularly enjoyed. Swift is no favorite of the very young, even through the veil of bowdlerization which surrounds the carefully expurgated editions to which they have access. But as a boy grows up, he finds the bitter draught of satire a strong and manly tonic, and I came to enjoy Swift, even though I shuddered as I read him. I also came to enjoy the milder but still vigorous vein of Thackeray, and to forgive or even to delight in his long-windedness. But above all satirists, I came to love the heart-rending cries of Heine, in which not one word is missing or in excess to obscure his love and his venom. I know, as did my father, almost every word of his *Hebraische Melodien,* and there are no poems that can move the Jew in me to greater pride or agony.

These books I read, not once, but many times over, lying

face down on my bed, and sucking the last savor out of the phrases I had scanned many times before. I have never been a great reader of new things; but what I have read and loved, I have taken into my memory, so that it has become a part of me, never to be discarded.

In the same way, I relearned my Latin and my Greek. The lapidary poetry of Horace is not merely something buried between the pages of my schoolbooks: it is engraved on the tablets of my memory. The sweep and grandeur of Homer are recollections which I can never forget. I may not be much of a classicist in the technical, competitive sense, but the roots of a classical education are firm in me.

At this time, my sister received a copy of Ruskin's *Modern Painters*. I read it avidly, and I thoroughly enjoyed Ruskin's rather academic drawings and the magnificent poetry of his language. Although I found his ventures into what might be called quasi-science dogmatic and incorrect, I could not help paying tribute to his superb talents as an observer. The book was my introduction to an appreciation of painting, sculpture, and architecture, but my later experience has taught me that though it is a brilliant commentary on the arts, there is a certain willfulness in Ruskin's prejudices, and that his study needs to be supplemented with a direct acquaintance with the great works of art and by a more catholic attitude toward the arts of the non-European countries.

The summer of 1912 we returned to the town of Sandwich and, to our delight, settled in the little interior valley at the foot of Flat Mountain and Sandwich Dome. The house we rented was known as the Tappan Place. Our next-door neighbors were the happy-go-lucky family of a Cambridge banker, ranging in ages from less than ten to the middle twenties. With one exception they were girls. They were great trampers, and with my own newly excited interest in climbing, we shared many quick jaunts up and down Sandwich Dome and Whiteface. I found the girls attractive, and was particularly taken by the one who was nearest my own age. Although I do not believe that I ever avowed my own admiration for her, there is probably still standing behind our house more than one beech tree that bears the marks of my jackknife.

I continued to tramp the woods with my father, but it was already clear that as my vigor increased, his had commenced to wane. The mixed pleasures of a heavy pack and a night on a balsam bed were no longer for him.

I had decided to work with Royce for my Ph.D. the next year in the field of mathematical logic. However, Royce's health had broken down, and Professor Karl Schmidt of Tufts College consented to take his place. Schmidt, who, I later learned, was a summer neighbor of ours in New Hampshire, was then a young man, vigorously interested in mathematical logic rather than the religious philosophy which later became his field during his tenure of office at Carlton College. Schmidt set me as a possible topic a comparison between the algebra of relatives of Schroeder and that of Whitehead and Russell. There was a lot of formal work to be done on this topic which I found easy; though later, when I came to study under Bertrand Russell in England, I learned that I had missed almost every issue of true philosophical significance. However, my material made an acceptable thesis, and it ultimately led me to the doctor's degree.

Schmidt was a patient and understanding teacher with Huntington's quality of being able to bring a young man to intellectual productivity by easy stages. If I had not had gentle handling that year, I do not believe that I could have come through it unmarked, for besides my dissertation, there were two ranks of dragons awaiting me.

The first mild fire-dragons were my topical examinations, which were written. Behind these loomed the fiercer dragons of my oral examinations. I passed through my topical examinations with my head bloody but unbowed. One incident in connection with them does not redound to my credit. All of us who had taken the examinations were very curious about our grades, and we found a compliant janitor who had access to the professors' rooms and the papers on which the grades lay. I regret to say that I teased him into telling me what these grades were, and I let at least one other candidate into the secret. It was a mere gesture of misjudged curiosity and misjudged good will, and bribery took no part in it, though I was accused of such a bribery later on.

I dreaded the oral examinations far more than the topicals. I went around to the houses of my various professors to take them. In every case the professor was kind and obliging, and in every case I went through the examination in a sort of trance scarcely understanding the words said to me. With Professor Woods, who examined me in Greek Philosophy, I found that I had forgotten almost every word of Greek that I had ever known, and was scarcely able to construe the simplest passage of Plato's *Republic*.

I must give my father every credit of seeing me through the great ordeal of the oral examinations. Every morning he went for a walk with me to keep up my physical condition and to reinforce my courage. Together we walked over many parts of Cambridge which were as yet unknown to me. He would ask me questions concerning the examinations that were ahead and would see to it that I had a fair idea as to how to answer them.

Nevertheless, at my own valuation, I should have failed every examination; but examining professors for the doctorate are likely to be more human and sympathetic with the student than the student is with himself, and to give him the benefit of every doubt. The terror of the student is familiar to all examining professors and is a part of the normal environment of the examination for which they automatically compensate,

A PHILOSOPHER DESPITE HIMSELF

so that no doctoral examination is taken at its full face value but is always interpreted in the light of such other data as the professor may have concerning the student's ability.

I have often been an examining professor myself in the course of my duties at the Massachusetts Institute of Technology. I have learned that terror deserves sympathy and is pardonable and indeed normal; and although the attempt to exert ingenuity on the spot and to lift oneself out of a difficulty while standing before the examiner is commendable, bluff is inexcusable. It is not the timid student but the glib yet ununderstanding student who has the most to fear.

After my oral examinations on specific subjects, I proceeded to the last part of the gantlet that I was to run, which consisted of my examination on my doctoral dissertation before the entire group of Harvard philosophers. This examination is theoretically the most critical stage in the ordeal of the candidate for the doctor's degree. But in fact no well-regulated department will permit a candidate to proceed so far unless it is substantially sure that he is going to pass. Furthermore, the candidate has now the tremendous advantage of reporting on a subject in which he is theoretically better informed than any of his examiners, so that there is no excuse for an honest man to be terrified, except by being tongue-tied and timid; and this, as I have said, is what his professors are most ready to condone. In fact, the oral examination for the doctorate is much more an examination in the student's conduct before a class than in anything else, and exerts a considerable amount of influence in the selection of those students who are going on to preferred academic positions. Here I do not believe that I did particularly well.

In the period between my examination and the last date duly allowed by Harvard statutes, I copied my dissertation on the typewriter in such a form that I might with decency embalm it in the Harvard archives. Some of this work was done at home, and some under the gaze of the portraits of former Harvard worthies in University Hall. Harvard did not at that time require the publication of a doctoral thesis. I am convinced that the Harvard attitude was sound, for it is unjust to accept the judgments of periodical editors as the criterion by which to give a university degree. Moreover, the forced private publication of such theses as do not meet the requirements of the editors of scientific journals places a heavy tax on the pocket of the student without doing any corresponding good to the profession at large. Privately published, these are universally inaccessible and, in general, little read. I am glad that the requirement of publication of the thesis is gradually disappearing.

It is often supposed that a man's doctoral thesis should be one of the best things he ever does, and should give the full measure of the man. I do not believe in this. A doctor's thesis is nothing but a specific piece of work by which a journeyman qualifies himself to become a master of his craft; and if he does not exceed this level a dozen times in the course of his career, he is a very poor master indeed. I know that many believe that the dissertation should stand out for years above a candidate's subsequent work, but this demand is often ignored in practice. It is only when a man has his dissertation behind him and is not pestered with the prospect of future formal requirements to fulfill that he can do his best work as a free man, with his task itself as the goal and not the spurious goal of a certain academic and social position. The thesis should be good, but if the scholar's work does not soon exceed the level of the thesis, the candidate is well on his way to becoming one of those desiccated homunculi you find in faculty meetings of our third-rank colleges.

If my own dissertation had been the only piece of scientific work I have ever produced, it would have been a most unsatisfactory ticket of entry to a career of learning. However, as the facts have developed, it did give me the training in the organization of scientific material which led me in the next two years to a series of papers that I should much prefer to represent my induction into a scholarly career.

I have known more than one student who has waited to present a thesis for his doctorate, even after he has produced a number of acceptable papers, until he can write that one paper that will allow him to break into print in the learned world with a maximum of impetus and *élan*. It is of course a fine thing if a youngster can establish himself as an important figure with his first work. Nevertheless, I feel that many a student has placed too much emphasis on this point, and has wasted years waiting for the great idea to come to him, which he might have devoted to experience in publishing and in receiving the public criticism of his printed work. It is altogether too much to expect to become a great man on the first try; and if the course of one's late work contains such material as need not be a matter of shame, it makes very little difference whether the first paper is excellent or barely conforms to the necessary standards for the doctorate.

Time hung heavy on my hands toward the end of the spring of 1913, after I had passed my doctoral examinations and while I was still waiting for Commencement and the apotheosis of hearing President Lowell declare that I was duly admitted to the society of learned men.

The year 1912-1913 was that of the demolition of the old Harvard library and preparation for the construction of the new building given by Widener. The old building had only been adapted for library purposes by a continuous series of improvisations and interior reconstructions, so that although it was probably one of the most authentic specimens of early American academic Gothic, its time was over and it had to go. There was a certain Roman holiday pleasure in seeing its pinnacles and vaults demolished by a great swinging iron ball; and indeed, the original workmen must have done the work so honestly that at times even this sort of violence made very little headway. The noise was intolerable, and our philosophical classes in Emerson Hall were conducted to an obligato of donkey engines and crumbling walls.

Yet with all the progress that the destruction of the old building signified, we felt that it marked the passing of an age. Nevermore would the library convey to us its anachronistic suggestions of medievalism, and the ample lawns about it were to be cut up forever by the crowding bulk of the Widener Library. My father always felt in view of his familiarity with the stacks, and the work that he did there, that the *d* in Widener might well be dispensed with. Be that as it may, the great convenience of the Widener Library as a storehouse for books never seemed to me to be matched by any particularly endearing qualities in its architecture. It was a cold and forbidding building, and later on during the war years the great stairway was decorated with two cold and forbidding paintings of the military might of America. They were Sargents, but they certainly weren't top Sargents.

At that time one of the places where I was most often to be found was the Harvard Philosophical Library. It was a pleasant place, and the librarian, Dr. Rand, was the exact Harvard equivalent of the slightly desiccated English don. He was an excellent historian and bibliographer, so that it was always interesting to search through the bookcases for something new and exciting. For example, we found that a number of the books which had been left to the library by William James were full of autographic marginal notations by James himself, which were rather less decorous than they would have been if they had originally been intended for public scrutiny. The James copies of the books of Royce and Bertrand Russell were particularly amusing. When Rand found out that the books were really priceless treasures, he locked them up in his private case and removed them from free public inspection.

Even after this there were many treasures which we found interesting reading. The English philosopher, F. C. S. Schiller, notwithstanding the lack of any great profundity on his part, had a pretty satirical vein which was always amusing to us. Then too, one never knew what was to be found among the newly arrived books and periodicals. Thus a casual leafing through these was part of our weekly routine.

During the course of the spring, I began to look at a number of educational periodicals in order to satisfy my curiosity about their treatment of the infant prodigy. This curiosity received a sharp punishment when I found, in a journal edited by G. Stanley Hall of Clark University, an article by a Miss Katherine Dolbear, the daughter of the late distinguished physicist of Tufts College.[1] This article was devoted to a discussion of Berle, Sidis, and me, case by case and name by name. Miss Dolbear was obviously not impressed by our record. She had presented with meticulous precision, not merely our official records at our several schools, but everything that she had been able to gather as to the opinion held of us by our undergraduate companions.

In no aspect was it a gratifying document. I had long been aware that my social development was far behind my intellectual progress, but I was mortified to find how much of a bore, boor, and nuisance Miss Dolbear's record made me out to be. I had thought that I was well on the way to the solution of my problems. Miss Dolbear's article made me feel like the player of parchesi whom an unfortunate cast of the dice has sent back to the beginning of the board.

I showed the article to my father, who was as furious as I had been humiliated. Father sent a letter of protest to be

[1] Katherine Dolbear, *Pedagogical Seminary*, Vol. 19, ''Precocious Children,'' p. 463.

published in the next number of the *Pedagogical Seminary,* although this did not serve any particular end. Our family lawyer was unable to give us much satisfaction in the matter. An attempt to seek a legal remedy would have subjected me to publicity far more dangerous and vicious than anything to which I had yet been exposed. Even in theory American law does not take great cognizance of the right of an individual to privacy, and a libel action, in order to be successful, must allege some specific damage as the result of the libel. Thus it is very dangerous to call a lawyer a shyster or a doctor a quack because such allegations do contribute very definitely to the injury of the professional standing of their object. However, I had as yet no profession; and although I hoped to have one in the future, damage to this would have been hard to prove and impossible to assess. This was the precise point that was raised later by the lawyers for *The New Yorker* in their defense of the suit instituted against them by W. J. Sidis; and the success of this defense confirms our judgment in not pushing the matter.

I regard this attitude toward libel eminently unjust. In the first place, to cast serious suspicions on a budding career seems to me an offense more serious than to interrupt one that is well under way. Secondly, an assault on the self-esteem of a person who is already in a difficult and questionable position is quite as great an injury as any physical assault could be. I think that a reasonable moral standard in such matters is exhibited by the practice of the medical journals, which I consider to be so well established that it would not be very difficult to give it the force of law. It is in the public interest that medical cases be reported precisely and freely in the professional journals. However, it is regarded as a grave offense to give the name of the patient or any data that could serve to identify him, at least without his voluntary and expressed consent. When a photograph of him is shown as part of the necessary documentation of the case, it is the custom, if the eyes and face do not show some essential part of the symptoms of the disease, to obliterate them in the print. I see no reason why a pedagogical journal, or indeed a journal without any scientific pretensions, should be given more latitude in such a matter. This is not a question of the freedom of the press, but it is eminently a question of the necessary correlate of such freedom: the responsibility of the press.

During my last year at Harvard I had applied for a traveling fellowship. Of course I was greatly excited when the news came that I had won it. Two places suggested themselves as alternate destinations: Cambridge, where Russell was then at the peak of his powers, and Turin, which was famous for the name of Peano. I learned that Peano's best days were over, and that Cambridge was the most suitable place for a training in mathematical logic. I then wrote to Russell, for it was necessary to obtain the permission of my teacher before I set out on my studies.

« XIV »

EMANCIPATION

Cambridge, June, 1913–April, 1914

WE RETURNED to New Hampshire that summer and I had a good chance to rest up for the year to come and to acquaint myself further with the mountain region. The mountains were an eternal delight to me. They are beautiful even now, but in those days before the war and the threat of war, before the extensive lumbering which the two World Wars called into being, before the motorcar and its reduction of distances to nothing and much of the roadside to a rural slum, the country was beautiful indeed. As one whose physical activity is somewhat limited by his increasing years and the vicissitudes of an active life, I look back with a certain sadness to a time when the mountainsides were as nothing to my efforts, and when twenty minutes of rapid striding would carry me to a bank of lacy wood sorrel. From this bank I could look up to the boles of mighty trees, each fit to be a mast of a king's ship. I felt a sense of romantic union with the hills and the forest.

One of my chief domestic tasks was to fetch the mail and

EMANCIPATION

the milk. Every day I tramped two miles to the little post office at Whiteface village and two miles back again, part of it with the milk pail digging into the palms of my hands. I was eager to go for the mail because the key to adventure awaited me there: my letter of acceptance from Russell.

Professor Huntington had recommended to me two mathematical books for summer reading before starting my work with Russell. They were Bôcher's *Modern Algebra* and Veblen and Young's *Projective Geometry*. The first book did not impress me so much then, although I have reread it many times and found it most useful as an introduction to matrix theory. The second book I took to my heart as the most consistent exposition of the postulational standpoint that I had found anywhere. I worked out almost all the problems of the first volume, which was the only one existing at the time. While the book had two authors, Young of Dartmouth was already somewhat invalided, and the personality of the book was chiefly that of Professor Oswald Veblen of Princeton. He was the founder of the great mathematical school of Princeton as well as the scientific founder of the Institute for Advanced Study, also at Princeton. He is without doubt one of the fathers of American mathematics.

The entire family was to go abroad for the winter. We had been tempted to go earlier, and had even gone so far as to make some negotiations for tickets, but this had been at the time of the Balkan Wars, and my father had considered the political heavens too stormy to risk it. Now, however, we really did embark. We had picked a boat of the Leyland Line, a small twig of the great I.M.M. combine, which ran cattle boats and carried a few passengers between Boston and Liverpool. I can remember that in those lucky days it was possible for fifty dollars to book a cabin for oneself and to have the run of the boat.

We left Cambridge by the subway and the East Boston

Tunnel for that desolate region of slums and docks known as East Boston. Here our ship was tied up. I remember that it hurt me to run across a maze of railroad tracks with heavy baggage in my hands, under the conflicting and self-contradictory orders of my father.

It was a heavenly relief to go aboard. The white-jacketed stewards served us biscuits and beef tea even before we had left the harbor. While we were still in the old familiar Boston harbor, with the Bunker Hill monument showing plain, we were already in a foreign territory: the manners of the stewards, the customs of eating and drinking, the very language that people spoke, were all new and strange to us.

My parents had maintained an almost instinctive position that the English they spoke and had learned was the only proper English language, and that all other forms of the speech had something illegitimate about them. I daresay Father would have made the adjustment to Basque or Tibetan more readily than to the change between the English of the American Boston and the English of London or of Lancashire.

For it was the English of Lancashire that predominated on the boat. It is a language that I have heard many times since; and although it is perhaps not the most beautiful of all forms of English, it has something of the winning quality of good bread and good cheese.

The passengers were few, and the radio bulletins of news of the world were not obviously obtrusive. The trip was long, uncrowded, and peaceful. The food was adequate but stodgy. There was nothing to watch except marbling waves, or the casual flirtation of an old sea captain's daughter with the wireless officer. With a little shuffleboard and a little chess, we made the trip very comfortably. And one morning we found ourselves tied up at the landing stage in the Mersey.

The formalities of landing were simple. It was a Sunday morning, and after we had bought our tickets for London, we had a meal of bread and cheese at a pub, and took off. I looked out of the train window, and renewed the impressions of the English countryside I had seen before as a child. In particular, I recalled the ivy, the smaller farms and fields, the brick and stone buildings, the less wooded landscape, and the seemingly smaller trees.

From Euston Station we made our way to Bloomsbury, which was then even more than now the natural barracks of the academic visitor of moderate means. We put up at a hotel at Southampton Row, which I recognized many years later as the scene of one of Graham Greene's more dismal tales of refugees and espionage. With the aid of our Baedeker, we found one or two possible vegetarian restaurants. We looked up father's old friend, Israel Zangwill in his lodgings in the Temple, and made plans for my stay in Cambridge. The rest of the family was to go to Munich for the winter. Constance was to study art and Bertha was to go to a private school for teen-age girls.

Father went up with me to Cambridge. We looked up Bertrand Russell in his rooms at Trinity, and he helped us to orient ourselves. While we were in Russell's rooms a young man came in whom my father took to be an undergraduate and who excited no particular attention in us. It was G. H. Hardy, the mathematician who was to have the greatest influence on me in later years.

It appeared that it was not necessary for me to matriculate inasmuch as Harvard and Cambridge had certain agreements concerning the privileges of advanced students. I therefore could not expect to live in college and it was necessary to find a landlady for me in town. My father did not spend much effort in placing me in lodgings. In one place he asked me in the presence of the landlady what I thought of the place. I was caught. I was forced to tell him on leaving that it seemed to me one of the most miserable, dirty, and inconvenient lodgings I had seen. Instead of canceling the word-of-mouth agreement that we had made, Father trusted to the improbability of my ever meeting the landlady again, and left matters to take care of themselves. He was in a hurry to catch the train back to London. Finally I was left *faute de mieux* with another slovenly little landlady in New Square. She had made some agreement to furnish me at the minimum price with the vegetables and cheese necessary for my vegetarian life.

It was at that time impossible for the American boy with anything like a normal bringing up to be completely free from a certain Anglophobia. The wars between the two countries, including the undeclared hostilities of our Civil War, were united with a certain latent enmity of tone in some of the English reviews in such a way as to comb a Yankee's hair the wrong way. More than all of these influences, the efforts of a few ardent American Anglophiles had the effect of making the American boy brandish the flag and let the eagle shriek.

Yet later, when I came back from England, I had learned that there was a very close and permanent bond between myself and England, and more especially between myself and Cambridge. I had learned that the English were very different from the Anglophiles in that, once one had penetrated the protective layer which they assumed against Americans and other foreigners, they were quite willing to admit that there were aspects of England in which God was not in his Heaven and something was definitely wrong with the world. I found that the English were as distrustful as I was of the Anglophiles' cure-all, which was to import English institutions to America, cut up into numbered pieces and wrapped in straw, as if they were Tudor manor houses. In short, I found that the England of the Anglophiles was a

EMANCIPATION

cloud-cuckoo-land existing neither on one side of the ocean nor on the other, but merely in the souls of the elect.

I came to find that among the institutions in which I had lived those which were most similar to English life and cast the most light upon it were in many cases the most specifically American institutions of my childhood. The country life of Ayer and Harvard, although it was a country life with neither squire nor established vicar, was a country life with very English roots. My New Hampshire farmer friends would probably have damned their opposite numbers in the Lake Country from here to Kingdom Come, and would have been received with similar objurgations; but despite the mutual hostile reserve and the difference in the dialects, the attitude would have been much the same on both sides. It would have taken only a few weeks of mutual contact for the one and the other to become aware that there was not terribly much difference between their attitudes or their presuppositions.

The England that I first saw was one which had not yet been shocked by World Wars and indeed which had remained at peace since the times of Napoleon, except for colonial wars and the major conflicts in the Crimea and South Africa. It was an England that was heaven for the rich and very close to hell for the poor. It was an England in which it was harder for a working man to become a scholar than it now is for a Mexican peon. This stratification and the snobbishness attendant on it—which was even more a masochism on the part of the poor than a sadism on the part of the rich—is something which, while some elements of it may remain, has passed out of the picture as completely as the France of the *ci-devants* did at the time of the French Revolution.

My landlady gave me my first introduction to the sort of English snobbishness and subservience that was then rife but which has since become much less common. She, a slovenly, mean little woman, did not approve of our neighbor two doors away. She said, "Ow, 'e's only a tridesman's son," even though the rank of tradesman was something vastly higher than any to which she could ever lay claim.

The university men of 1913 were young sprouts of the aristocracy, or at least of a well-established middle class. Since then I have seen the rise of the subsidized undergraduate. The working class boy, stunted by undernourishment in his early childhood and in the womb of his mother, with bad teeth and horny hands, wearing a hand-me-down suit and big clumping boots, has come to be supported by exhibitions and scholarships through his primary and secondary schools and his university. These are the men I now know as young dons; accepted because of their ability and character, but often cursed with a social awkwardness which they have had to unlearn with a very genuine and conscious effort. More than one of them has confided to me the pains which he had to take at the beginning to develop a good line of high-table back-chat.

The phenomenon of which I speak is spread far beyond the cloistered courts of England's university. It is a relief to me now to be able to sit on a park bench and talk with an English workingman who will neither resent me as a "toff" nor whine for some advantage. Indeed, to the present generation of Englishmen who read this book it may seem that I am accusing their predecessors of vices that are so far from their own make-up that the newer Englishman is unable to conceive them. But I can say that as I have revisited England year by year, I have seen servility decline and a universal manliness and comradeship come to the front.

So much for my reminiscent view of Cambridge. At the time of my first arrival, after spending a few days learning the lay of the land, I was hopelessly and utterly lonely. Term had not yet begun, so that there was no chance for me to make new acquaintances. I wandered about the colleges and

EMANCIPATION

in the Backs, and the utter beauty of the buildings and the foliage was more than a little solace to my nostalgia. Meanwhile I met one or two undergraduates: a Hindu who lodged in the same house as I and a young Englishman two houses away. They both belonged to St. Catherine's College, and they invited me to participate in the meetings of a discussion club belonging to that college.

I have no specific recollection of what was said and done at that club at St. Catherine's. I remember that I was asked to read a paper and to say a few words. I did so, and I have a dim memory that I covered myself with shame and confusion. I certainly spent the first few weeks in Cambridge in learning the English point of view, and in sloughing off some of the most impermissible of my many awkwardnesses. I know that my callow nationalism got me into more than one childish quarrel.

Nevertheless, I feel that this was a critical period in my formation, and that I owe a great deal of gratitude both to my teachers and to my undergraduate friends of those days. I found in them a receptiveness and a tolerance of ideas which had not been characteristic of Harvard, and a challenging dialectical skill in presenting them.

Although I had very good times with several of the young undergraduates in their clubs and social groups and "squashes" and at tea in their rooms, there was a group of slightly older men on the boundary between the undergraduate and the don who were particularly kind and helpful to me. One of them was F. C. Bartlett, now Sir Frederic Bartlett, and a professor of psychology at Cambridge University. My impression is that he had come from one of the more modern English universities, and that at that time his prospects for a career were not particularly bright. I found his steady quietness and his refusal to be stampeded by any argument a healthy tonic for my own impulsiveness. His criticism was always fair and not to be bribed by friendliness. I am glad that our relations have been kept up over these many decades, and that the basis for them has not changed in any essential way.

Bernard Muscio was another one of my seniors who was very kind to me and who helped me to grow up. He was born in Australia, where he had obtained his first degree. His alertness and quickness of reaction made him an important figure in the Moral Science Club, better known as the Moral Stinks Club, and more than once we joined forces in a dialectic assault on those with whom we did not agree.

Two of my early associates of a very different kind were C. K. Ogden and I. A. Richards. Ogden, who had succeeded in prolonging an undergraduate career over an unheard-of period of years, then lived above a gateway in Petty Cury, where his rooms were adorned with photographs of practically every important man in intellectual England. Among the manifold facets of his being, he was a journalist and he solicited from me an article which he published in the *Cambridge Magazine*, and whose nature I have entirely forgotten in the course of these many years. Richards and he were close companions, and I believe that during my stay in Cambridge the collaboration which later led to the publication of *The Meaning of Meaning* had already begun. At any rate, their interest in semantics was manifest.

One of the things that most impressed me at Cambridge was the rather too cloistered atmosphere of the English university scholar. He had come from a school devoted to the needs of adolescence, which constituted the most essential and characteristic part of his education, to a university built according to a scheme closely paralleling that of his adolescence. If he were successful, a career was open for him for his whole life under much the same auspices.

The English universities, although they were no longer

EMANCIPATION

exclusively the celibate, clerical institutions which they had been in the earlier nineteenth century, still retained much of their monkish character. Thus the young man going into mathematics carried into his valuation of mathematical work a great deal of the adolescent "play-the-game" attitude which he had learned on the cricket field. This, although it contained much of good, and led to a devotion to scholarship difficult to find in our more worldly life, was not conducive to a fully mature attitude toward his own work.

When G. H. Hardy—as the reader may easily find in his book, *A Mathematician's Apology*—values number theory precisely for its lack of practical application, he is not fully facing the moral problem of the mathematician. It takes courage indeed to defy the demands of the world and to give up the fleshpots of Egypt for the intellectual asceticism of the pure mathematician, who will have no truck with the military and commercial assessment of mathematics by the world at large. Nevertheless, this is pure escapism in a generation in which mathematics has become a strong drug for the changing of science and the world we live in rather than a mild narcotic to be indulged by lotus eaters.

When I returned to Cambridge as a mature mathematician after working with engineers for many years, Hardy used to claim that the engineering phraseology of much of my mathematical work was a humbug, and that I had employed it to curry favor with my engineering friends at the Massachusetts Institute of Technology. He thought that I was really a pure mathematician in disguise, and that these other aspects of my work were superficial. This, in fact, has not been the case. The very same ideas that may be employed in that Limbo of the Sages known as number theory are potent tools in the study of the telegraph and the telephone and the radio. No matter how innocent he may be in his inner soul and in his motivation, the effective mathematician is likely to be a powerful factor in changing the face of society. Thus he is really dangerous as a potential armorer of the new scientific war of the future. He may hate this, but he does less than his full duty if he does not face these facts.

In laying out my course, Russell had suggested to me the quite reasonable idea that a man who was going to specialize in mathematical logic and in the philosophy of mathematics might just as well know something of mathematics. Accordingly, I took at various times a number of mathematical courses, including one by Baker, one by Hardy, one by Littlewood, and one by Mercer. I did not continue Baker's course long, as I was ill prepared for it. Hardy's course, however, was a revelation to me. He proceeded from the first principles of mathematical logic, by way of the theory of assemblages, the theory of the Lebesgue integral, and the general theory of functions of a real variable, to the theorem of Cauchy and to an acceptable logical basis for the theory of functions of a complex variable. In content it covered much the same ground that I had already covered with Hutchison of Cornell, but with an attention to rigor which left me none of the doubts that had hindered my understanding of the earlier courses. In all my years of listening to lectures in mathematics, I have never heard the equal of Hardy for clarity, for interest, or for intellectual power. If I am to claim any man as my master in my mathematical training, it must be G. H. Hardy.

It was while I sat in this course that I wrote the first mathematical paper which I saw in print. Looking back on this paper, I do not think it was particularly good. It was on a reordering of the positive integers in well-ordered series of large ordinal numbers. Still, it gave me my first taste of printer's ink, and this is a powerful stimulant for a rising young scholar. It appeared in the *Messenger of Mathematics*, which was published in Cambridge, and I had the satisfaction of seeing it through press on the spot.

EMANCIPATION

I attended two courses of Bertrand Russell. One was an extremely elegant presentation of his views on sense data, and the other a reading course on the *Principia Mathematica*. In the first course I could not find myself able to accept his views on the ultimate nature of sense data as the raw material for experience. I have always considered sense data as constructs, negative constructs, indeed, in a direction diametrically opposite to that of the Platonic ideas, but equally constructs that are far removed from unworked-on raw sense experience. Apart from our disagreement on this particular matter, I found the course new and tremendously stimulating. In particular, I found myself introduced to Einstein's relativity, and to the new emphasis on the observer which had already revolutionized physics in Einstein's hands and which was to revolutionize it even more completely in the hands of Heisenberg, Bohr, and Schrödinger.

There were only three of us in Russell's reading course, so that we made rapid progress. For the first time I became fully conscious of the logical theory of types and of the deep philosophical considerations which it represented. I became shamefully aware of the shortcomings of my own doctoral thesis. Nevertheless, in connection with the course I did one little piece of work which I later published; and although it excited neither any particular approval on the part of Russell nor any great interest at the time, the paper which I wrote on the reduction of the theory of relations to the theory of classes has come to occupy a certain modest permanent position in mathematical logic. It was published soon after I was nineteen in the *Proceedings of the Cambridge Philosophical Society*, and this paper represents my true introduction into mathematical thinking and writing.

It is not very easy for me even at this distance to write of my contact with Bertrand Russell and of the work I did under him. My New England puritanism clashed with his philosophical defense of libertinism. There is a great deal in common between the libertine who feels the philosophic compulsion to grin and be polite while another libertine is making away with the affections of his wife and the Spartan boy who concealed the stolen fox under his cloak and had to keep a straight face when the fox was biting him. This does not endear the philosophical libertine to me. The old-fashioned rake had at least the fun of don't-care; the puritan is working within a code of known restrictions which tends to keep him out of trouble. The philosophical rake is as bound as the puritan, and has to steer a course in as narrow a channel; but it is a channel which is poorly lighted and poorly buoyed. I expressed myself very freely in this matter, and I am quite certain that Russell heard my comments to a friend one dark night when we met on the street as we were returning to his quarters. Though he never gave a sign of hearing me, this experience rendered me particularly apprehensive of his criticism.

I know that Russell regarded my Harvard thesis as inadequate, in that I did not enter sufficiently into the problem of logical types and into the paradoxes that mark the difficulties of establishing a fundamental postulational system for logic, as opposed to a derived postulational system for a specific construction with a recognized logic. As for myself, I already then felt that an attempt to state all the assumptions of a logical system, including the assumptions by which these could be put together to produce new conclusions, was bound to be incomplete. It appeared to me that any attempt to form a complete logic had to fall back on unstated but real human habits of manipulation. To attempt to embalm such a system in a completely adequate phraseology seemed to me to raise the paradoxes of type in their worst possible form. I believe I said something to this effect in a philosophical paper which later appeared in the *Journal of Philosophy, Psychology and*

EMANCIPATION

Scientific Method. Bertrand Russell and the other philosophers of the time used to term this journal "the Whited Sepulchre," an allusion to the simple white paper cover in which it appeared. My heresies of that time have been confirmed by the later work of Gödel, who has shown that within any system of logical postulates there are questions that cannot receive a positive answer through these postulates. That is, if one answer to these is consistent with the original postulates, it can be proved that the opposite answer is equally consistent with them. This treatment of the problem of decision has rendered obsolete a considerable part of the task undertaken by Whitehead and Russell in the *Principia Mathematica*.

Thus logic has had to pull in its horns. The limited logic which remains has become more nearly a natural history of what is in fact necessary for the consistent working of a system of deduction than a normative account of how it should be worked. Now, the step from a system of deduction to a deductive machine is short. The *calculus ratiocinator* of Leibnitz merely needs to have an engine put into it to become a *machina ratiocinatrix*. The first step in this direction is to proceed from the calculus to a system of ideal reasoning machines, and this was taken several years ago by Turing. Mr. Turing is now occupied in the actual construction of computing and logical machines, and has thus completed a further step in the direction of the *machina ratiocinatrix*. The remarkable thing is that I myself, quite independently of him, have recently also taken the step from my early logical work to the study of the logic of machines, and have thus again met the ideas of Mr. Turing.

To go back to my student days with Russell, although there were many points of disagreement and even of friction, I benefited enormously by them. His presentation of the *Principia* was delightfully clear; and our small class was able to get the most out of it. His general lectures on philosophy were also masterpieces of their kind. Besides his consciousness of Einstein's importance, Russell also saw the present and future significance of electron theory, and he urged me to study it, even though it was very difficult for me at that time, in view of my inadequate preparation in physics. I do not recall, however, that he was quite as explicit and accurate in his valuation of the coming importance of quantum theory. It must be remembered that the epoch-making work of Nils Bohr was very new at the time, and that in its original form, it did not lend itself particularly to a philosophical interpretation. It was only some twelve years later, in 1925, that the conflicting currents aroused by the earlier work by Bohr began to be resolved and that the ideas of De Broglie, Born, Heisenberg, and Schrödinger showed that quantum theory was to mark as great a revolution in the philosophical presuppositions of physics as had the work of Einstein.

On the social side, the most distinctive aspect of my contact with Bertrand Russell lay in his Thursday evening parties, or as they were called in view of the number of guests, his "squashes." A very distinguished group of men foregathered there. There was Hardy, the mathematician. There was Lowes Dickinson, the author of *Letters from John Chinaman* and *A Modern Symposium*, and the bulwark of the liberal political opinion of the time. There was Santayana, who had left Harvard for good to take up his residence in Europe. Besides these, Russell himself was always an interesting talker. We heard much of his friends, Joseph Conrad and John Galsworthy.

Three of the most important moral science dons with whom I came in contact, all fellows of Trinity, were known as the Mad Tea Party of Trinity. Their identities were unmistakable. It is impossible to describe Bertrand Russell except by saying that he looks like the Mad Hatter. He has always been a very distinguished, aristocratic Mad Hatter, and he now is a white-haired Mad Hatter. But the caricature of Tenniel almost argues an anticipation on the part of the artist, even though I am told that the original of Lewis Carroll's description and Tenniel's caricature was an actual hatter at Oxford, and that his "Anglo-Saxon Attitudes" were really the effect of an industrial mercury poisoning. McTaggart, a Hegelian and the Dr. Codger of Wells's *New Machiavelli,* with his pudgy hands, his innocent, sleepy air, and his sidelong walk, could only be the Dormouse.

The third, Dr. G. E. Moore, was a perfect March Hare. His gown was always covered with chalk, his cap was in rags or missing, and his hair was a tangle which had never known the brush within man's memory. Its order and repose were not improved by an irascible habit of running his hand through it. He would go across town to his class, with no more formal footwear than his bedroom slippers, and the space between these and his trousers (which were several inches too short) was filled with wrinkled white socks. He had the peculiar habit of emphasizing his words on the blackboard by running them through with chalk-marks instead of underlining them. He used to make the most withering remarks in philosophical discussion, in a breathless but smiling and unperturbed manner. "Now really," he would say, "you can't expect any sane person to hold a view like *that*!" On at least one occasion at a meeting of the Moral Science Club, he brought to a state of tears Miss E. E. C. Jones, the Mistress of Girton, lovingly known as "Mammy Jones" to the unregenerate. Yet when I came to know him and to depend on his criticism of my work, I found him kind and friendly.

There is among the dons a premium on individuality which often becomes a premium on eccentricity. I have been told by some of my Cambridge friends that they thought that certain of my less conventional habits had been adopted with a view of acquiring acclaim. At any rate, the fact is there; and while I do not think that Russell's mannerisms (which were very slight) were any more than a genuine manifestation of his aristocratic background, I am quite certain that the untidiness of G. E. Moore and the academic unpracticalness of McTaggart had been cultivated very carefully. They had the flavor of a crusty old port—a flavor that does not reach its full perfection without the expert intervention of the cellarman.

During the term I made quite a number of acquaintances, and my mantelpiece was adorned with the cards of discussion clubs. I had an invitation to visit some friends of Mr. Zangwill, who lived about fifteen miles out in the country; and I turned up there, dusty and bedraggled, after walking the entire distance. In general, by the end of the term I was finding my social place in Cambridge. I had even begun to have a certain fondness for my new environment.

Yet most of the time I was desperately uncomfortable in a physical sense. My landlady had been paid little enough; and yet that could hardly excuse the raw carrots and inedible Brussels sprouts which she gave me in lieu of proper vegetarian meals. I eked out my diet with occasional penny bars of chocolate and the like, but the net result was that I was half starved.

In my leisure hours, and I had many of them, the Union and its library were my salvation. My membership in the Harvard Union had enabled me to make use of the facilities of its Cambridge counterpart, and I even took part in one or two of the famous undergraduate debates. Moreover, some of my friends occasionally asked me to dine at the Union, so that I learned something of the amenities of an English club.

I found the Cambridge environment far more sympathetic to me than I had found that of Harvard. Cambridge was devoted to the intellect. The pretense of a lack of interest in intellectual matters which had been a *sine qua non* of the life of the respectable Harvard scholar was only a convention and an interesting game at Cambridge, where the point was to work as hard as you could in private while pretending to exhibit a superior indifference. Furthermore, Harvard has always hated the eccentric and the individual, while, as I have said, in Cambridge eccentricity is so highly valued that those who do not really possess it are forced to assume it for the sake of appearances.

Thus when the beginning of December came and I left to spend the Christmas holiday with my family in Munich, I was both happier and more of a man than I had ever been. The trip was a lark. I crossed to the Continent by the Harwich route, and did not have a bad passage. I was up well before dawn to see the lights of the Hook of Holland, and I was pleasantly bewildered by hearing the Dutch speech of the porters. I breakfasted in the big, empty, echoing railway station, and dawn saw me well on the way toward Rotterdam. I don't know by what use of English, bad German, and gestures I persuaded a porter to take my trunks across town in a barrow to another station, but I soon found myself bound for Cologne, uncomfortably seated in a third-class compartment, all windows hermetically sealed, in an atmosphere that seemed to be made partly of commercial travelers and partly of tobacco smoke.

I arrived in Cologne in the early afternoon and found quarters for myself in a very cheap hotel, which I now believe to have been nothing more than a Kellnerheim. There was no way to get to Munich that day, so I took a walk about the town and tried to correlate my impressions with my memories of my childhood trip more than eleven years earlier. I found that in fact there was a good deal that I did remember: for example, the station, the bridge, and the cathedral.

I went to Munich the next day in a through carriage. I was delighted with everything that I saw on the way, from the forests with touches of snow on them to the villages and stations which looked to me like the illustrations accompanying the set of Anker building blocks I had played with as a child. My German was as yet insufficient to enable me to communicate with my fellow travelers, so I spent most of my attention on the landscape outside. The scenery along the Rhine was a reawakening of my memories of the former trip of my boyhood, and the wooded mountains of Franconia had not a little suggestion for me of the White Mountains.

My family met me at the Munich station and took me to the old-fashioned but centrally placed apartment that they had rented. Although the apartment house had long been invading America, I had never lived in one up to that time, and the apartment-house mode of life was to my parents something altogether undesirable. Indeed, I had been brought up to regard the city life of the apartment as a deprivation and a misfortune for the people who had to resort to it. The fact that our landlady spoke no English, and that my mother was not confident of her German, did not ease matters.

My father spent his time working at the Bavarian Court and National Library. Away from the Harvard library (where by long experience he could put his hand on every book he wanted), and under the usual restriction of exclusion from the stacks and the eternal pinpricks of the abominable system of cataloguing which was then standard outside the United States, his work languished. Moreover he was disappointed that his name was less well known by his European colleagues than he had expected, and that he had few personal contacts with them or none at all. To some extent this was only to have been expected, for my father had a very individual way of working, and he had no hesitation in contradicting flatly the presuppositions of the scholars of his time and in writing in a blunt manner which was an affront to their sense of self-importance. Germany was then a hierarchical form of society, from the workingman at the bottom to the Kaiser on top; and within the greater frame the university people were a lesser hierarchy. For a mere foreigner with no place in this system to beard a whole school of learned German *Geheimrats* was a scandal beyond words. My father, who was a most sensitive man, could not fail to be aware of the atmosphere about him.

Until that time, my father had always been a great admirer of German culture and German education. Although he had resented the militarism and officialism that had developed since his own youth, he was fundamentally a German liberal of the middle of the last century. His Russian Tolstoyism was an influence that ran parallel to the German influences in his development and did not contradict them. He had for a long time looked forward to the period when he should return to Germany and be accepted as a great scholar within the German frame of things. When this did not occur, and he found himself rejected, or perhaps merely not accepted, this emotional longing turned to a hate which was as bitter as only a hate for a lost love can be.

My sisters had been properly placed in the appropriate schools. I do not remember through what vicissitudes of attempts at musical and artistic training my sister Constance had gone before she decided to work at the Kunstgewerbeschule, or school for industrial art. Bertha was placed at a fashionable and respectable girls' school, the Institut Savaète, where she made good progress in her general education and her understanding of things German. I do not remember quite how we disposed of Fritz's school time.

By now I had become sufficiently grown up to be a fairly acceptable comrade for my father. We went together to various lectures and beer-hall meetings where interesting subjects were being discussed. I remember one such meeting on international peace and understanding at which the speaker was David Starr Jordan, the famous ichthyologist and president of Stanford University. I remember drinking my glass of beer and feeling very much the man among the German students.

My parents took me occasionally on their outings to the Plätzl and other cabarets; and I often went with my sisters to the movies, which were just beginning to give signs of their later development. There was also a small amount of visiting fairs and historical museums. However, my chief delight was the Deutsches Museum: a museum of science, engineering, and industry. Part of the exhibits were historical and old-fashioned; but the museum led the world in its demonstration of the technique of scientific experiments, which the visitor could actually work for himself behind the protecting glass cases by pulling strings or by turning knobs. There were some delightful old attendants there who were ready to put themselves at the service of the interested visitor, and to show him particular tidbits not always brought to the attention of the general public. I remember one in particular who put himself out to be nice to me; he possessed a few words of English and a most delightful Bavarian brogue.

The Deutsches Museum had an extremely modern scientific library; there I read assiduously the various works that Russell had assigned to me. I remember among them the original papers of Einstein. I have said that Russell was among the first philosophers to recognize the overwhelming importance of Einstein's work in that *annus mirabilis* 1905, in which he had originated the theory of relativity, solved the problem of the Brownian motion, and developed the quantum theory of photoelectricity.

Another delight of that vacation was the Englischer Garten, even in its snow-covered winter state. I remember the skaters on the pond near the Chinese Pagoda. I was not aware at the time that the Englischer Garten was laid out after the plans of a New England Yankee from Woburn, Massachusetts, the

EMANCIPATION

great and disagreeable Benjamin Thompson, Count Rumford, and paymaster for Benedict Arnold.

I returned to Cambridge in January. I felt myself already much more at home in the town, and much less lonely. I continued to distribute my time between philosophy and mathematics, and began a second paper for the Cambridge Philosophical Society. This time I tried to use the language of the *Principia Mathematica* to describe series of qualities, such as those found in the color pyramid, which escaped from a treatment of series given by Whitehead and Russell because they were not infinitely extensible in both directions. What I found necessary was a logical treatment of systems of measurement in the presence of thresholds between observations whose differences were barely noticeable. In the paper I used certain ideas related to those of Professor Whitehead, who was then at the University of London and who had recently employed a new method of defining logical entities as constructs from entities of a primitive system possessing no particular specific properties rather than as the objects of a system of postulates. I wrote to Professor Whitehead for an appointment and visited him in his house in Chelsea, where I met the whole family. Little did I think at that time that Professor Whitehead was to end his long and useful career as my neighbor at Harvard University, and that as a very inept pupil of his daughter I was later to learn some of the rudiments of the art of rockclimbing in the crags of the Blue Hills and in the Quincy quarries.

I had intended to complete the year in Cambridge, but I found that Russell had been invited to Harvard for the second semester and that therefore I would be marking time in Cambridge during the May term. At Russell's own advice, I decided to finish the year at Göttingen, studying mathematics with Hilbert and Landau and philosophy with Husserl. I returned to Munich for the vacation between the last two terms. My father had already left for the United States, where he was consoled in his loneliness by the companionship of some younger colleagues in the German department, but my mother and the rest of the family were still in Munich. During the year I had read that Harvard offered a number of prizes for essays by students, both undergraduate and graduate. I found that I was eligible to compete for one of the Bowdoin Prizes and submitted a rather skeptical essay, which I called "The Highest Good." It was intended as refutation, or at any rate as a denial, of all absolute ethical standards. Bartlett did not think much of it, either as a composition or as a philosophical essay, but at any rate it won one of the prizes. I am quite sure that Sir Frederic still regards this as a shortcoming of Harvard rather than as a success of my own.

My departure from England was marred by a very unpleasant experience with my landlady. When my father had made the arrangements with her, he had thought that he was committing me for a single term or less. However, by the custom of Cambridge the term is of a certain specified length, which is longer than the period known as full term, during which the students are supposed to be in residence, and all lodgings contracts are or were made for the longer period. As the second term drew to the end, I found my landlady insisting on this contract. From being demanding she became pressing, and from being pressing she became insulting. I retorted in kind, which made the matter worse. Some of my undergraduate friends with whom I consulted suggested a minor riot at my landlady's expense; but although I was foolish, I was not quite so foolish as that. When I tried to take one of my trunks out of the house on my own back, the landlady impounded the other; and when I asked the police to help me in getting back my own property, they told me that it was a civil matter and that they could have nothing to do with it.

I had been living on an absolutely minimal sum, so that

when I paid the landlady the sum necessary to redeem my trunk, I found that I did not have enough money left to get to Munich. I borrowed a small sum from the hall porter of the Union. Out of shame I borrowed too small an amount. The result was that when on the train down to Munich, I had to decide whether to have a cheese sandwich for breakfast and go hungry for lunch, or vice versa. I do not remember which way I decided it. The upshot of it all was that I landed in Munich with not a single coin in my pocket. Luckily the check room charges were paid on reclaiming the baggage, so I left my baggage at the station and walked over to our apartment.

I found matters rather in a crisis at the apartment. The smoldering friction between the landlady and my mother had burst into flame, now that my father was no longer there to help with his German. Mother went house-hunting, and after much effort we managed to find an apartment in the suburbs, well out toward the northern end of the Englischer Garten, and almost abutting on it. Here we were completely at peace.

« X V »

A TRAVELING SCHOLAR IN WARTIME

1914–1915

AFTER STAYING in Munich a few weeks, I went on to Göttingen before the beginning of term, to take part in a psychological congress which was being held there and to see my old friend Elliott, the Harvard psychologist, who had come there to participate in it. I have not much recollection of the congress, but I found the town a delightful medieval gem, with the circuit of its old walls almost complete.

For what seemed to me a ridiculously small sum, I immatriculated at the university and began to search for a lodging and for a vegetarian restaurant at which to eat my meals. I found my lodging just outside the walls at the house of a Fräulein Büschen. It was a half-timbered villa in Swiss style, and my room, though dark, was adequate. Fräulein Büschen, who had been a music teacher, attended to the business side of the establishment. She managed this very competently, and left the matter of our breakfast and other domestic needs to her sister, who did not aspire to the social distinction claimed by the music teacher. Somewhere around the establishment there was an obscure brother, who had been trained as a dentist but who did not seem to practice that art.

I remember at least one party the Büschens gave for their student lodgers, which was attended by some nice girls from the neighborhood. I particularly remember the shock to my 1914 New England susceptibilities when I found that the whole company, both men and women, were smoking cigarettes and were not averse to drinking to the point of becoming slightly tipsy.

I found my vegetarian restaurant in the Theaterstrasse, in the home of a Frau Bauer. She was a widow with a considerable number of daughters of varying ages, who assisted her and the cook. The girls waited barefoot at the table, for vegetarianism was not the only respect in which the Bauers departed from the norm. They were clothes-reform people, youth-movement people, health faddists, and anti-Semites as well. It was in their restaurant that I first saw that vile sheet *Hammer*, which even at that epoch already contained all the lies and blasphemies that Hitler and Goebbels disseminated so disastrously at a later period.

Bigoted as they were, the Bauers were not entirely bad. Their food was good and cheap, and they were personally amiable enough. They served an oatmeal preparation with the uninviting name of *Haferschleimsuppe,* or "oat slime soup." It was inexpensive and filling.

I often wonder whether the poor Bauers ever realized what serpents they had taken to their bosom in the person of myself and a young Scottish mathematical physicist by the name of Hyman Levy. Levy, who is now a distinguished professor at the Imperial College of Science and Technology in London, notwithstanding his impeccable Glasgow accent, was, like myself, just what his name purported him to be; and yet we two Jewish sons of perdition defied the wrath of the anti-Semitic periodicals about us and ate, nay even enjoyed, the cheap and savory meals prepared for us. When I think that in addition to his Jewishness, Professor Levy has become a bulwark of the Left in English politics, I imagine that old mother Bauer, if indeed she be dead, is turning cartwheels in her grave.

Service was very slow in the Bauer restaurant. We used to take our plates out to the cook in the kitchen and have her fill them for us from the pots on the stove. This informality was possible because we were a happy-go-lucky, impecunious mixture of Germans and Americans, Britons and Russians, and the low prices attracted many of us who had no fads or dietary interest at all. We used to read our papers there.

The *Kneipen* or drinking bouts of the German students are well known. We too—the English and the Americans—had our *Kneipen* at the meetings of two separate but commensal societies known as the British and the American Colonies. The heads of these two colonies were called the Patriarchs, and Levy was the British Patriarch. The two clubs occupied a room above the Franziskaner Restaurant. The beer supply was steady and unfailing, and the floor was on such a slant that navigation was difficult, even without a cargo of beer. We shared a piano, a *Kommersbuch*, or German students' song book, and a Scottish Students' Song Book, which was Levy's personal property. Our meetings were long, moist, and harmonious. We paid our respects to the land that had welcomed us as well as to the lands that had fostered us by singing indiscriminately in English and in German. We were the rowdiest *Kneipe* in town and had been compelled to leave two or three former quarters by the protest of the proprietor or of the police.

There was one member whose name I shall not mention out of respect to such of his kinsmen as may be alive, though no contemporary Göttinger can ever forget him. It was not Early, though that is what I shall call him. Early was the son of an American publisher of hymn books, and he seemed deter-

A TRAVELING SCHOLAR IN WARTIME

mined to live down his family's good name. He was married, and his wife and young daughter had the full sympathy of the united colonies. Early had looked around the world for a soft spot to settle, and he had picked Göttingen. In some vague way he had managed to immatriculate at the university, where he had been a student for about ten years, though I had never heard of his taking or attending a course. When any of the American students made a pilgrimage to a nearby city for questionable purposes, it was Early who was their guide, friend, and philosopher.

These activities, I must confess, were only minor manifestations of his personality. The serious purpose of his life was drink. Never did a meeting of the united colonies come to an end but he was drunk as a coot, and some one of us had to see him safely home. I believe he was courteously apologetic on these occasions; and indeed there were about him some curious remnants of a man of breeding.

When I next went to Göttingen in 1925, Early was gone, though his fame had not vanished. I am told that he stayed on well into the First World War but that before we entered that conflict his family had been able to bring him home. In view of his age at that time and of his manner of life, he must long be dead by now. Yet his type will never die; and wherever scholars come together and there is a comfortable life, there will be the blight of the perpetual student. I write this chapter in a room in a hotel just off the Boulevard St. Germain in Paris. Around the corner at this moment dozens of Earlys at the Cafés Flore and Deux Magots are sipping their *apéritifs* and trying to turn the more serious young people into their own likenesses.

I had a varied range of acquaintances in Göttingen. I remember a student who had come from the imperial Russian police and who was studying psychology in connection with his professional career. Another of my philosophical acquaintances was a very bright Russian Jewish boy. On one occasion several of us were at a small party at the house of the landlord of the latter fellow—a retired chief forester with the bluntness that one naturally associates with this profession. I don't remember all the things we discussed, but my philosophic friend asked me to say something about the work of Bertrand Russell. After I had spoken my little piece, my fellow student brought out: "But he doesn't belong to any school."

This was a serious shock to me: that a philosopher should be judged not by the internal implications of his own work but by the company he keeps. It was not, in fact, the first time I had heard of this intellectual gregariousness, which was common in Germany of those times but not confined to it, but I had never really run up against a first-class example of that sort of pedantry. I had, it is true, encountered the collective manifesto of the American New Realists. But the weakness of this group was so apparent that it seemed to me their solidarity and mutual mental support resembled that of a group of college students returning home after an exciting evening following a great football game: they literally could not stand up alone.

However, when I now saw a similar phenomenon in Germany, it claimed to be more than a protective huddle. The implication was that the privilege of a man to think depended on his having the right friends. Later, when I was to come back to the United States, I found that I myself had the wrong friends. I had studied with great men but they were not the men on the American scene. The Harvard department of mathematics would have none of me because I had learned the greater part of my mathematics at Cambridge and at Göttingen. When the new Princeton department recruited its men after the war, I was already too much of a lone wolf to be welcomed there. It is true that these two universities (together with Chicago) never reached the extremes of some of the German universities in their corporate isolationism, but they have made a good try at it.

I attended a course on group theory given by Professor Landau and a course on differential equations under the great Hilbert. At a later period, when I had become more familiar both with the literature of mathematics and with the techniques of mathematical research, I came to a clearer understanding of these two men. Hilbert was the one really universal genius of mathematics whom I have met. His excursions from number theory to algebra and from integral equations to the foundations of mathematics covered the greater part of known mathematics. There was in his work a complete grasp of tools and techniques; however, he never put into the background the fundamental ideas behind these. He was not so much the manipulative expert as the great mind of mathematics, and his work was comprehensive because his vision was comprehensive. He almost never depended on a mere trick.

Landau, on the other hand, was the chess player *manqué*. He believed in presenting mathematics as a sequence of propositions analogous to moves on a chessboard, and he did not believe in the nonsymbolically expressible part of mathematics that constitutes the ideas and strategy behind the moves. He did not believe in mathematical style, and as a consequence his books, effective as they are, read like a Sears-Roebuck catalogue.

It is interesting to contrast them with the work of Hardy, of Littlewood, or Harald Bohr, all of whom wrote in the manner of cultivated and mature men. Landau, on the other hand, had intelligence, but he had neither taste nor judgment nor philosophical reflection.

It is impossible to mention the Göttingen of those times without referring to Felix Klein, but for one reason or another I did not meet him the semester I was there. I rather believe that he was out of town or in ill health. When I later met him in 1925, I found him very much of an invalid indeed: a grave, bearded man with a blanket over his knees, who sat in his magnificent study and discussed the mathematics of the past as if he were the Muse of mathematical history herself. He was a great mathematician, but by this time in his career he had become rather the *Geheimrat*, the elder statesman of mathematics, than the producer of mathematical ideas. There was something kingly about him which suggested to the career men of American mathematics that they, too, might be kings if they followed in his footsteps, and they treasured his little mannerisms (such as the way he speared his cigar with a penknife) as if by a careful observation of this ritual they might charm their way to greatness. Many years later I became aware that two generations of Harvard mathematicians had learned this trick from him.

Besides these mathematical courses, I sat in Professor Husserl's course on Kant and his seminar on phenomenonology. The philosophical courses left very little impression on me, as my German was inadequate for the subtleties of the philosophical language. I got something at the time from the mathematics courses, but much more by that sort of intellectual doubletake that allows one to realize at a later date the importance of what one has already heard but not understood.

Even more important for my intellectual training than the courses were the mathematical reading room and the Mathematical Society. The reading room contained not only what was probably the most complete collection of mathematical books in the world but also the reprints that Felix Klein had been receiving over the years. It was a great experience to browse among the books and the reprints.

The Mathematical Society used to meet in a seminar room, where the tables were covered with the latest numbers of all the mathematical periodicals of the world. Hilbert would preside, and professors and advanced students sat together.

Papers were read by students and professors alike, and the discussion was free and incisive.

After the meeting we would traipse across the town to Rohn's café in a beautiful park at the top of a hill overlooking the town. There we would have a mild glass of beer or coffee, and would discuss all sorts of mathematical ideas, both our own and those we had learned in the literature. There I came to know the younger men, such as Felix Bernstein, who had done some remarkable work in Cantor theory, and little Otto Szasz, with his high-heeled shoes and his bristling red mustache. Szasz was my particular crony and protector, and I am very happy that later when the Hitler regime came in, I was able to help place him in the United States.

The combination of science and social life in the *Nachsitzungen* at Rohn's café up the hill was particularly attractive to me. The meetings had a certain resemblance to those of the Harvard Mathematical Society, but the older mathematicians were greater, the younger men were abler and more enthusiastic, and the contacts were freer. The Harvard Mathematical Society meetings were to the Göttingen meetings as near beer is to a deep draft of Münchener.

About this time I had my first experience of the concentrated passionate work that is necessary for new research. I had the idea that a method I had already used to obtain a series of higher logical type from an unspecified system could be used to establish something to replace the postulational treatment for a wide class of systems. The idea occurred to me to generalize the notions of transitivity and permutability, which had already been employed in the theory of series, to systems of a larger number of dimensions. I lived with this idea for a week, leaving my work only for an occasional bite of black bread and Tilsiter cheese, which I bought at a delicatessen store. I soon became aware that I had something good; but the unresolved ideas were a positive torture to me until I had finally written them down and got them out of my system. The resulting paper, which I entitled *Studies in Synthetic Logic*, was one of the best early pieces of research which I had done. It appeared later in the *Proceedings of the Cambridge Philosophical Society* and served as the basis for the Docent Lectures which I gave at Harvard about a year afterward.

Mathematics is too arduous and uninviting a field to appeal to those to whom it does not give great rewards. These rewards are of exactly the same character as those of the artist. To see a difficult, uncompromising material take living shape and meaning is to be Pygmalion, whether the material is stone or hard, stonelike logic. To see meaning and understanding come where there has been no meaning and no understanding is to share the work of a demiurge. No amount of technical correctness and no amount of labor can replace this creative moment, whether in the life of a mathematician or in that of a painter or musician. Bound up with it is a judgment of values, quite parallel to the judgment of values that belongs to the painter or the musician. Neither the artist nor the mathematician may be able to tell you what constitutes the difference between a significant piece of work and an inflated trifle; but if he is never able to recognize this in his own heart, he is no artist and no mathematician.

Granted an urge to create, one creates with what one has. With me, the particular assets that I have found useful are a memory of a rather wide scope and great permanence and a free-flowing, kaleidoscope-like train of imagination which more or less by itself gives me a consecutive view of the possibilities of a fairly complicated intellectual situation. The great strain on the memory in mathematical work is for me not so much the retention of a vast mass of fact in the literature as of the simultaneous aspects of the particular problem on which I have been working and of the conversion of my fleeting impressions into something permanent enough to have a place in memory. For I have found that if I have been able to cram all my past ideas of what the problem really involves into a single comprehensive impression, the problem is more than half solved. What remains to be done is very often the casting aside of those aspects of the group of ideas that are not germane to the solution of the problem. This rejection of the irrelevant and purification of the relevant I can do best at moments in which I have a minimum of outside impressions. Very often these moments seem to arise on waking up; but probably this really means that sometime during the night I have undergone the process of deconfusion which is necessary to establish my ideas. I am quite certain that at least a part of this process can take place during what one would ordinarily describe as sleep, and in the form of a dream. It is probably more usual for it to take place in the so-called hypnoidal state in which one is awaiting sleep, and it is closely associated with those hypnagogic images which have some of the sensory solidity of hallucinations but which, unlike hallucinations, may be manipulated more or less at the will of the subject. The usefulness of these images is that in a situation in which the main ideas are not yet sufficiently differentiated to make recourse to symbolism easy and natural, they furnish a sort of improvised symbolism which may carry one through the stages until an ordinary symbolism becomes possible and appropriate. Indeed, I have found that there are other mental elements that may readily lend themselves for preliminary symbolic use in the solidification of ideas in mathematics. On one occasion during a bout with pneumonia, I was delirious and in considerable pain. But the hallucinations of my delirium and the vague reactions of pain became associated in my mind with some of the difficulties yet hounding me in an incompletely solved problem. I identified my suffering with the very real malaise that one feels when a group of ideas should fit together and yet cannot be brought together. However, this very identification gave me sufficient markers for my problem to enable me to make some real progress in it during my illness.

But life in Göttingen was not all research for me. I needed outdoor exercise, and I took my tramps with my English and American colleagues in the woods south of Göttingen and in the region of Hanover-Münden. My favorite lunch tidbit may seem rather indigestible, but it was cool and delightful: a Tilsiter cheese sandwich, a dill pickle, a glass of lager beer, and a raspberry ice.

There were many interesting things to see in Göttingen. There was a fair on the Walkenmühlenwiese, near our favorite swimminghole in the river Leine; and we were delighted to see the side shows and to hear the barkers of just the sort I had known at a New England carnival, against this unfamiliar background. I remember the different sorts of beer which I furtively tasted at the local Automat, and the bathhouse with its different classes of baths and its ample towels of the more expensive grades. I remember the two-hour classes, and the little buffet at which we bought sandwiches and Leibnitz-Keks in the fifteen-minute interval between the two halves.

The summer term was drawing towards a close, and the coming storm of the First World War made itself known in the papers by the heat lightning of the assassination at Sarajevo. The diplomatic ineptitudes that followed did not relieve the tension. Luckily, I had planned to return to America, and I had already secured my third-class passage on a Hamburg-American steamer.

I derived many lasting benefits from Göttingen. My contact with the philosophers was not very satisfactory. I do not have the type of philosophical mind that feels at home in abstractions unless a ready bridge is made from these to the concrete observations or computations of some field of science. From the mathematicians I also got relatively little in the formal courses. Landau's group theory course was a hard-driving plunge through a mass of detail with which I was not fully prepared to cope. I was able to follow Hilbert's course in differential equations only in parts, but these parts left on me a tremendous impression of their scientific power and intelligence. It was much more the meetings of the Mathematische Gesellaschaft which taught me that mathematics was not only a subject to be done in the study but one to be discussed and lived with.

Besides this, at Göttingen I learned to meet people both like me and different and to get on with them. This marked an important step ahead in my social development. The net result was that I left Germany much more a citizen of the world than when I first went there. I can say this very truly, even though I had not been fond of all aspects of the Göttingen environment and although in the war that immediately followed, I passed through a definitely anti-German phase. Yet when I went back to Germany during the confused years between 1919 and Hitler, for all the alienation I may have felt on political issues, there was a large intellectual element in Germany with which I had a sufficient common basis of past experience to feel myself a part.

My year at Cornell and my two years of graduate study in philosophy at Harvard had represented the continuance of my adolescence and my gradual introduction to independent research. They were satisfactory as far as my purely intellectual progress went, but they did not see me clearly out of the Slough of Despond. I was quite as aware as those about me that the way of the infant prodigy is beset with traps and snares, and while I knew perfectly well that my purely intellectual powers were above the average I knew equally well that I was to be judged by standards according to which a moderate degree of success would take on the appearance of failure. Thus I did not escape the floundering that generally goes with adolescence; and although this floundering was at a far higher intellectual level than that of the majority of teenagers, it represented a more than usually severe and doubtful struggle with the forces of uncertainty and of my own inadequacy.

It was the year in Cambridge and in Göttingen, however, that gave me my emancipation. For the first time I was able to compare myself intellectually with those who were not too much above me in age and who represented in fact the cream of the intellectual crop of Europe and even of the world. I was also subject to the inspection of first-rate men like Hardy and Russell and Moore, who could see me without the glamour of my precocity and without the condemnation which belonged to my epoch of confusion. I do not know whether I was outstandingly brilliant in their eyes, but at least they (or some of them) regarded my career as a reasonable bet. I was not under the immediate tutelage of my father and did not have to weigh myself in his somewhat loaded scales. In short, I had been initiated into the great world of international science, and it did not seem utterly hopeless that I should accomplish something there.

I was learning all the time how to comport myself as a social being and what the requirements were for living among those of other traditions and customs. My study in Germany represented an even greater break with my childhood, and an even greater necessity for me to adapt myself to foreign standards or at least not to come into a head-on collision with them.

The Sarajevo pot gradually went from a simmer to a boil. By the time I had got to Hamburg, there were posters on the street, calling on all Austrians subject to military service to return to their country. The city was full, and the *Christliches Hospiz* at which I stayed could only put me in a bathroom in the waiters' home which was their annex. I heard singing in the street at night and thought the war had come;

but it had not, and I spent the time before dawn walking around the outer Alster basin.

I took the train for Cuxhafen, where I embarked on the *Cincinnati* of the Hamburg America Line. A day and a half later I saw the mobilization of the British fleet in Spithead, and about two days afterward we received the news that Germany and England were at war and that the radio station was closed. While we were bound for Boston, we did not know whether we should be able to make it, and at one time there was some talk that we might be sailing for the Azores. However, the sun showed that this was not the case, and we made Boston as intended. This ship was then laid up at a Boston dock until the United States entered the war, when it was taken over as an American transport and later torpedoed by the Germans.

My father met me at the boat, much relieved to find me safe and sound. We took the train together up to New Hampshire. I noticed that Father treated me with more respect than he had ever done: more as a grown man. We talked about the war during the train trip. I was surprised to find how definitely my father's opinion and the university opinion he represented had crystallized against Germany.

The war news was bad. We had hoped for a quick ending of the war, but the German line bit deeper and deeper into Flanders and France, and even when it was held by the taxicab army of Marshal Joffre, it was clear that we were in for a long, desperate, and uncertain war of positions. It was then the children of my generation knew that we had been born too late or—barely possibly—too soon. Santa Claus died in 1914. We surmised that life was to be such a nightmare as Kafka has since described, from which one awakes only to become aware that the nightmare is real, or from which one awakes into an even worse nightmare.

I had written to Bertrand Russell to inquire whether it was

advisable to return to Cambridge on the new and augmented traveling fellowship which Harvard had granted me for the academic year 1914–15. He wrote that it would be safe and desirable, and I booked passage from New York on an old ship of the American Line. It dated from the days of auxiliary sail, and had a yacht bow and a bowsprit. It seemed enormously romantic to me.

My two aunts, who had by now risen in the clothing trade world and spent much of their time in Paris, saw me off at New York. The trip was slow but agreeable. There were young people aboard with me who tried to forget the war. We played a version of golf with shuffleboard sticks and disks, chalking the holes on the deck, and using ventilators, cleats, and deckhouses as hazards. There was an elderly couple from Australia who watched our antics benignly. They ran a sort of agricultural school when they were at home. I was later to see them in the grim London of the war, where their cordiality was a great consolation to me.

Thus I arrived in wartime Cambridge. The air was heavily overlaid with gloom. Part of the Backs had been turned into an improvised hospital for wounded soldiers. In all the vacant spaces of the university, ugly shacks were springing up, of a temporariness more devastatingly permanent than any intended permanency.

In the Union there were lists of casualties, and distressed fathers and brothers were reading them in the hope that they might not contain the names of their own kinsfolk, yet with the expectation that sooner or later these kinsfolk would appear on them. *Blackwood's Magazine* contained monthly installments of Ian Hay Beith's book *The First Hundred Thousand*, which brought to us a sense of the immediacy of the war and of a certain participation in it.

The news continued black and ominous. My friends and colleagues were scarcely able to take their intellectual work with full seriousness, and the blacked-out streets and the white painted curbs added to a general feeling of gloom and doom. Finally, we began to hear suggestions that the Germans were soon to undertake a great unrestricted submarine campaign against merchant and passenger ships.

It was embarrassing for me to meet the soldiers everywhere, in the movies, in the streets, and even in the classrooms of the university, and to think that as a foreigner I was immune to the universal sacrifice. Several times I thought of enlisting, but was deterred by the fact that after all it was not yet my war and that to go into it before my parents were ready to accept the situation would be in some sense a very serious disloyalty to them. Then too, with my poor eyesight, I was not exactly the best soldier material; nor did I desire to sacrifice my life for a cause concerning the merits of which I was not yet fully convinced. Though I definitely preferred the English and French side of the war, I had not yet been excited to that righteous pitch of indignation to which my father had been carried by a complex of emotions which I have already described.

I was impressed with that freemasonry which exists among the British ruling class, whatever their political opinions, which makes it possible to share the intimate knowledge of many secrets carefully concealed from the press and the public. My attention was called to this very forcibly during this second stay in Cambridge. I was in the habit of receiving from my parents copies of the old Boston *Transcript*, that former fount of ultrarespectability and of reasonably accurate news. In one number I read of the sinking of the British cruiser *Audacious*. I had not seen a word about the matter in my British newspaper. I went to Russell with the news, and he told me that it had been common knowledge since it occurred, and that the *Illustrated London News* had published a photograph of the ship with the caption, *"An Audacious Picture!"* and no explanation of the audacity.

Furthermore, Russell seemed to be well informed about every other detail of the war which had been concealed from the public at large. Yet, at this time, Russell was an extremely unpopular figure with the officials of the British government. He was a conscientious objector and a pronounced pacifist; and when later America entered the war, he expressed himself concerning the American government in such hostile terms that he was sent to jail and ultimately deprived of his position at Cambridge.

To me this combination of being officially on the black books of the government and still personally in a position to receive from his official opponents information that was refused to the public at large seemed a remarkable tribute both to the stability of England and to the assured position of its ruling class at that time.

By Christmas I could stand the gloom of Cambridge no more and I went down to London. I found rooms in a melancholy turning off Holborn and spent much of my time reading my landlady's books about old London and checking up the places mentioned as they were to be found at the present time. I found my Australian friends in a Bloomsbury hotel. I looked up another Harvard fellow in philosophy, T. S. Eliot, who, I believe, had taken Oxford to himself as I had taken Cambridge to myself. I found him in a Bloomsbury lodging, and we had a not too hilarious Christmas dinner together in one of the larger Lyons restaurants. I also looked up the Whiteheads, and found that war bereavement had already struck them.

Some time after I returned I received a telegram from my parents telling me that the submarine threat was growing worse, and that I should come home on the first possible boat. Actually, Cambridge had almost closed down, and there was very little point in my staying any longer. I decided to finish my year at Columbia, so I booked a passage from Liverpool to New York and ultimately made a gloomy train journey to Liverpool under depressing war conditions. My companions in the compartment were a group of soldiers absent without leave who jumped off the train at the first stop before we arrived at Lime Street and into the arms of the military police.

I have been on winter trips across the Atlantic which have been as calm and as agreeable as any trip in the full summer time, but this March voyage was not one of them. The old ship took it green up to the bridge, and nothing but a youthful resiliency kept me from being sick all the way. Among my fellow passengers the most interesting were a family of Belgian refugees who were leaving a temporary asylum at Cambridge, England, for a more permanent one at Harvard University in the American Cambridge. Professor Dupriez was a distinguished professor of Roman Law at Lovain, and a charming gentleman, but he was also an impractical little scholar of the European type. The practical brains of the family and its energy belonged to his wife, a queenly and downright Flemish lady. There were also four children, two boys and two girls, who were too young to have completely lost the adventurous pleasure of the journey in the depression of defeat and exile. We were to see much of this family in the next few years.

The boat arrived in New York, and I was met by my New York kinsfolk. I went to Boston for a few days to pick up the continuity of our family life, and then I returned to New York to finish my year's fellowship at Columbia University.

I found the skyscraper dormitories of Columbia depressing after Cambridge and Göttingen. I also found the life of the place unsatisfactory in its lack of coherence and unity. Almost the only bond between the professors, who lived widely scattered in University Heights apartment houses or in suburban bungalows, was an almost universal antagonism toward Nicholas Murray Butler and everything that he stood for.

I did not get along too well with the other men in the dormitories. There was no intellectual bond between us, and I seem to have been completely lacking in tact. I insisted on criticizing my seniors intellectually in a way not becoming a boy of the age of a college sophomore. I threw my weight about giving bits of information which were not welcome to the men about me who were predominantly graduate students. It is true that I did not always know that these bits of information were particularly recondite and unwelcome. I intruded upon bridge games between an established set of cronies without making sure that I was welcome. I should have been far more sensitive to the response that my conduct evoked. They used to pester me by setting fire to the newspaper I was reading, and by other such feats of heavy-handed buffoonery.

Following the advice of Bertrand Russell, I studied with John Dewey. I also took courses with some of the other philosophers. In particular, I listened to lectures by one of the New Realists, but I was only able to confirm my impression of an undigested mass of the verbiage of mathematical logic, completely uncombined with any knowledge of what it was all about.

My term at Columbia was a makeshift at best and although I began to develop the intellectual consequences of my own ideas, I did not get much help from my professors. Indeed, the only one of them who was a great name comparable to those I had learned to appreciate at Cambridge and Göttingen was John Dewey; and I do not think I got the best of John Dewey. He was always word-minded rather than science-minded: that is, his social dicta did not translate easily into the precise scientific terms and mathematical symbolism into which I had been inducted in England and Germany. As a very young man I appreciated the help and discipline of a rigid logic and a mathematical symbolism.

About the time that I returned to America, I was told that the philosophy department at Harvard would look out for me for the next year with an assistant's position, and that I would be allowed to give a free series of Docent Lectures which were the prerogative at that time of every Harvard Ph.D. Hence I began to prepare for my Harvard Docent Lectures.

My research in New York was devoted to an attempt to establish a postulational and constructive treatment of *analysis situs* within the ideational and terminological frame of Russell's and Whitehead's *Principia Mathematica*. This was in 1915, many years before Alexander, Lefschetz, Veblen, and others had succeeded in doing very much the same sort of thing that I then tried to do. I had pages and pages covered with formulae, and I made a certain very real progress, but I was disappointed because I considered the bulk of results which I obtained distressingly small in view of the large structure of presuppositions which I had set up to obtain them; consequently I never carried my investigations sufficiently far to put them into publishable form. In neglecting to do so I may very well have lost the chance to be one of the founders of what has become a most fashionable mathematical subject. However, my early start in mathematical logic, a subject at which many mathematicians arrived only after a mature consideration of other fields, rather cloyed me with abstraction for abstraction's sake, and gave me a somewhat exacting sense of the need for a proper balance between mathematical apparatus and the results obtained before I have been able to consider a mathematical theory as intellectually satisfying. This has led me more than once to discard a preoccupation with a theory which I myself had at least partly originated, and which, because of the ease with which it has furnished doctoral theses, has become a fashionable field of studies. Here I refer in particular to the study of Banach's spaces, which I discovered independently of Banach in the summer of 1920, only a few months after his own original work and before its publication.

In this connection, let me say that the fact that I came from a field of the most abstract theory has always led me to put a great value on richness of intellectual structure and on the applicability of mathematical ideas to scientific and engineering problems. I have always had, and I still have, great suspicions and reservations in regard to work that is thin and facile; and until we had the correcting influence of the sense for application required by war work, I cannot deny that a large part of specifically American work and not a little of that found abroad have suffered from a certain thinness of texture.

I used to explore on foot the whole of Manhattan Island, even as far as the Battery. I went for walks on the Palisades near the Jersey side of the river with Professor Kasner of the mathematics department. He then lived in a part of Harlem at the foot of University Heights, before Harlem had come to have its present significance. Kasner would tell me much of his ideas on differential geometry, and he was a pleasant walking companion, who knew a much wilder Palisades region than can be found at the present time.

My stay in New York also marked my introduction to the American Mathematical Society, and my first visual acquaintance with most of the elder scholars of the group. At that time the hotel headquarters was that pile of Gay Nineties respectability, the old Murray Hill Hotel. The society was then more of a New York institution than it is at the present time, for it had indeed been founded by a New York group, and had been known for some years as the New York Mathematical

Society. There attached to it a little of a beer-hall flavor, which has evaporated with time and the increased prosperity and respectability of the scientist.

I spent Sunday and sometimes Saturday with my grandmother and other New York relatives somewhere up in the Spuyten Duyvil region of Manhattan. My relatives were very kind to me but I found the *gefüllte-fish* atmosphere of a north Manhattan apartment house a little stifling. On one occasion I ventured to accept the invitation of my cousin Olga to take a walk in the country with her and to visit a few friends. I should have given my grandmother this time, as she was old and diabetic and not likely to live longer than about a year. However, the indignation with which my mother received the news of my dereliction from duty was only in part due to her affection for my grandmother. It was at least in part due to her fear of my acceptance of Jewish environment in the more threatening form provided by Olga and the younger generation.

My mother's father died while I was at Columbia, and I saw my mother on the hurried trip through New York to Baltimore. Not long after that I received a telegram from my father summoning me home for an immediate conference. Short of money as I was always at that time, I rushed to the train and sat the night out in a coach. When I came home the awful news was broken to me. One of my former fellow students, who was an instructor at Harvard, had told the authorities of the philosophy department, who were considering my future career, that just before obtaining the doctor's degree I had bribed the janitor to show me the results of some of the examinations. I have already mentioned this incident and although my conduct was certainly not justifiable, bribery played no part in it. My father took me immediately to Professor Perry's office to confront me with my accuser, and I had the pleasure for once in my life to hear my father's magnificent repertory of invective applied, not to me, but to an enemy. The incident ended in my formal acquittal, but it did no good to me in my later search for a permanent job.

My stay at Columbia had not represented any part of my original plans for that year, and had been forced on me by the exigencies of the war and of my parental fears. It probably represents the low point of my academic career between the summit of my European trips and the gradual ascent by which I climbed to a position as a teacher and an independent research man. Here, if it appears thin, it is because it was in fact thin. Still, it taught me something about New York and something about academic life at a large city university. I had done a piece of my own scientific work which would have been important if I had at that time possessed the courage to see its originality and to bet upon it in the face of a general lack of interest in the new Analysis Situs.

« X V I »

TRIAL RUN:
TEACHING AT HARVARD AND
THE UNIVERSITY OF MAINE

1915–1917

WE RETURNED to New Hampshire for that summer. Raphael Demos, whom I had already known as a student in the Harvard philosophy department, and two other young Greeks, Aristides Evangelos Phoutrides of the Harvard classics department and Bouyoucos of the School of Agriculture at the University of Michigan, met me there in midsummer, and we set out on a mountain-climbing excursion.

It was a long, wilderness trip, the first that I had undertaken without the guidance of my father. Besides the beauty of the landscape and the pleasure of now being taken on a tramping trip by my contemporaries, the trip was enlivened by the personality of Phoutrides, who was a poet in his own right and a reader of much of the best in modern Greek poetry. It was a great revelation to see our own White Mountains through the eyes of a man who had formed his mountaineering skills on Parnassus and Olympus, and associated his pleasure in the sport of it with the genuine tradition of the classics.

Not long after this, we returned to the city and to the duties of term time. I acted as a section instructor in a large freshman course in philosophy, as well as instructor in my own right in a course in logic. In my undergraduate teaching I had to keep a record of attendance, read papers for one or two professors, and conduct a section at Harvard and a section at Radcliffe on elementary philosophy. I was in fact about the same age as most of my students in these two sections. I don't think I had very much stage fright in teaching, but it took a certain amount of courage to stand up before a crowd, particularly a crowd of girls about my own age, and hold the discussion to a progressive and orderly course. I don't know how I got away with it, because, as is natural with insecure beginners, I was far more pontifical then at the age of twenty-one than I am now at the age of fifty-eight. Still, Harvard and Radcliffe classes are traditionally tractable and I was interested in my subject and always ready for a dialectical fight.

I was grateful for the gift of gab. It is easier to prune a youthful loquacity and excessiveness than to cultivate the art of saying something when the words won't come. Besides, I was protected by my very inexperience from the awareness of what a show I was making of myself. This awareness came later, the next year, when I left the amenities of Cambridge for contact with the raw facts of life in the woods at Orono, Maine. Then I was to pay severely for my ineptitudes in classes that delighted to show up my powerlessness at discipline.

In addition to teaching, I had somewhat special duties. Professor Hattori of Tokyo was giving a series of courses on Chinese and Japanese culture and philosophy, and he needed the help of a young American in the routine work of taking attendance and of grading. I took on the job and found that it stimulated my interest in the civilization of the Orient, to which I had already been led in dealing with the problem of my Jewish origin through my interest in the general problem of undervalued peoples. This interest in the Orient was reinforced by the fact that my particular crony of that year, with whom I used to go tramping in the Middlesex Fells and the Blue Hills, was Chao Yuen Ren, a brilliant young Chinese, who had left graduate study in physics at Cornell to study philosophy at Harvard and who was equally versed in phonetics and in the study of Chinese music. Chao has continued to be my close friend all these years, and in the periods in which I have been unable to see him, I have received the *Green Letters* by which he solved the problem of his very extensive correspondence. These are printed documents of considerable length and merit by which he keeps his friends *au fait* with his affairs.

I may say in passing that Chao Yuen Ren has become perhaps the greatest philologist of Chinese and one of the two leading reformers of the Chinese language. He was Bertrand Russell's interpreter in China; and he married a charming Chinese woman doctor who is also the greatest interpreter of Chinese cookery to the West. They have four daughters, of whom the oldest two were born in the United States and are now married, and who aided their father during the late war in teaching Chinese at Harvard.

These friendships made me very conscious of the role of the non-European scholar at American universities. I was living through a period in which great changes were taking place in the relative and the absolute role of America in world science. Properly speaking, these changes were only a part of the general process by which countries have come up and gone down in their creative activities. This has since been evidenced by the way in which the overwhelming primacy of Germany had been reduced by emigration, war, and hardship. What I then found and still find more striking and significant were the changes in lands previously foreign to European culture,

such as China, Japan, and India, and in the new colonial
countries. Many of these have come within my lifetime to take
a major share in the Western intellectual world.

Besides my regular courses, I also gave what was called a
Docent course in constructive logic. For a number of years,
Harvard granted to every Harvard Ph.D. the right to carry
on a series of lectures in his chosen field without pay and
without authority for the students to count them for a degree.
Otherwise, the lectures were officially recognized by Harvard
University. I have already mentioned that it was my intention
to supplement postulational methods by a process according
to which the entities of mathematics should be constructions
of higher logical type, formed in such a manner that they
should automatically have certain desired logical and structural properties. The idea had an element of soundness, but
there were certain difficulties that I had not foreseen and
evaluated, which depended on the fundamental finiteness of
scope of our experience. My work was closely connected with
Bertrand Russell's notion of perspectives, and I suspect that
both pieces of work shared many of the same advantages and
disadvantages.

At that time a star of the first magnitude had appeared on
the firmament of the Harvard mathematics department. It was
G. D. Birkhoff. In 1912 Birkhoff, then twenty-eight, had
astonished the mathematical world by solving an important
problem in dynamical topology which had been put—but never
solved—by Poincaré. What was even more remarkable was
that Birkhoff had done his work in the United States without
the benefit of any foreign training whatever. Before 1912 it
had been considered indispensable for any young American
mathematician of promise to complete his training abroad.
Birkhoff marks the beginning of the autonomous maturity of
American mathematics.

He had continued his work on dynamics of the sort that

Poincaré had previously discussed, and was giving a course on the problem of three bodies. I enrolled in the course, but whether it was due to my insufficient preparation or to Birkhoff's rather difficult expository style, or in all probability to both, I made heavy weather of it and was unable to continue.

Both Birkhoff and Münsterberg were among the auditors of my Docent course. As the war went on, Münsterberg found his position at Harvard more and more difficult. He took the German side, whereas most of his colleagues, including my father, took the Allied side. Finally, Münsterberg wrote to my father a letter which my parents took to be insulting, and there followed an active quarrel in which Münsterberg alluded to his interest in my work and to his attendance at my course and his support of it. Naturally, the situation could not have been more embarrassing to me, and I showed more loyalty than tact in taking up the cause of my father.

While I had occasionally attended the Harvard Mathematical Society during my previous stays at Harvard, I now began for the first time to attend it regularly. It was an institution formalized in the typical Harvard way. The professors sat in the front row and condescended to the students in a graciously Olympian manner. Perhaps the most conspicuous figure was W. F. Osgood, with his bald oval head and his bushy divaricating beard, spearing his cigar with his knife after the pattern of Felix Klein, and holding it with a pose of conscious preciosity.

Osgood typified Harvard mathematics in my mind. Like so many American scholars who had visited Germany at the beginning of the century, he had come home with a German wife and German mores. Let me add that there is much to be said for marrying German wives—I am happy to have done so myself. In Osgood's time this New England pseudo-Germanism had been the fashion of the day. His admiration of all things German led him to write his book on the theory of functions in an almost correct German. There was no question that he had been impressed with the status of the German *Geheimrat*, and that he had longed to model academic life in America into such a form that he could imagine himself in such a position. He had done able work in analysis, against the resistance of those inhibitions that perpetually drive a certain type of New Englander from the original into the derivative and the conventional. Some of his ideas should have led him to the discovery of the Lebesgue integral, but he had not brought himself to the final step which might have led him to accept the striking consequences of his own conception. He must have had some rankling awareness of how he had missed the boat, for in his later years he would never allow any student of his to make use of the Lebesgue methods.

Another representative of the German period of American mathematical education was Professor Maxime Bôcher. He was the son of a former teacher of French, but he had been educated in Germany and had married a German wife. As in the case of Osgood, German was the language of his house; but in other ways he departed from the Osgood pattern. His work was more original and on a broader basis, and his personality was free from easily traceable mannerisms.

Of the other professors of mathematics perhaps the two that impressed me most were Professors Edward Vermilye Huntington and Julian Lowell Coolidge. I have already spoken of Huntington, whose very originality had collided with his Harvard career. He had been relegated to the Lawrence Scientific School and to the teaching of engineers, although his ability was greater in the direction of pure mathematics and logical ingenuity. He has lived to see his heresy become an orthodoxy, and today the postulational method attracts even more than its due share of Ph.D. candidates. He was an excellent, inspiring, and patient teacher.

Under the Lowell regimen, a kinsman of Lowell himself

and a descendant of Jefferson, such as Julian Lowell Coolidge, was bound to be among the favored. Coolidge had been educated in England and in Germany. His work was in geometry, in which he had shown great diligence and industry. With his wit, he had managed to synthesize an amusing personality, and he had succeeded in turning an inability to pronounce the letter *r* into a rather attractive individual trait.

To give a paper before the Mathematics Club constituted a real training in skill of presentation and in logic. Originality and power were not at a premium. Power in mathematics consists in possessing tools, conventional or otherwise, which enable one to solve a large part of the previously unsolved problems which one encounters in the course of one's work. It is the ability to create or to develop methods to match the demands of the problem—and it was not in high esteem in that environment. There was at that time no place where the interests of the advanced scholar beyond the early graduate-student stage were paramount, although such an organization has developed since in the form of the Mathematical Colloquium, which now has taken over a large part of the functions of the Mathematics Club.

For my physical exercise I combined hiking with a certain amount of wrestling in the gymnasium. Wrestling is one of the sports in which a myope may participate without unduly serious physical disadvantage. I never was really good at wrestling, but I was heavy and strong, and particularly strong about the shoulders, so that I gave better wrestlers a certain amount of exercise in banging me about. For a while I was a mass of boils and mat burns, in the best tradition of the sport.

When I returned from Cambridge and Columbia, I returned to an atmosphere of authority in family discipline which was not much less intense and all-pervading than it had been in my student days. Still there was one difference: I was no longer my father's pupil in any subject. The old idea of the family structure remained, but this time it hinged on the fact that I was now a breadwinner in the full sense and that I had begun to have a certain stature in my own right. Yet it was not until much later after my marriage that I can say I ceased to be in my father's eyes fundamentally the child from whom obedience was demanded.

From the time I returned from Europe, my parents had been accustomed to having Sunday teas for my father's students. My own students were also invited as well as the fellow students of my sisters. The Sunday teas of the professor with adolescent and postadolescent children is indeed a hoary tradition; nor has it changed its original intention. When I read in Thackeray of the Professor of Phlebotomy at Cambridge and of his attempts to introduce his daughters to promising undergraduates, it strikes an echo in my own memory. Nevertheless, I should be the last person in the world to poke fun at these teas, for they did in fact furnish me with the basis for a social life when other bases were scant, and my sisters as well as I first met our future spouses on these occasions. I learned much about how to handle myself socially, and I began to develop friends and acquaintances.

My father was Professor of Russian at Harvard, and it fell to his lot to extend courtesies to Russian visitors. During the war there were a great number of them, first on missions of varying importance involving the Russian government and later the refugees from the threatening storm of revolution or from the revolution itself. These were men and women of varying caliber. Some of them were on serious missions, such as winding up a purchasing campaign made on behalf of the Tsarist government. Some were there for the sake of their own hides and very little else, and these young elegants came to our teas, played Russian songs on our piano, and philandered all over our house. Even among these, there were some who had sufficient serious ability to make good in the new land; but there were also those whose connection with life was about as tenuous as that of froth with beer. For a while my mother and father were charmed with the social position of being hosts for this group of aristocrats, and my parents would compare their courtesy, sophistication, and *savoir-faire* with my own gaucherie, much to my disadvantage; but I was perfectly conscious all the time that our home was nothing but a backdrop to the amorous ballet of these delicate souls, and that they regarded us with utter indifference if not with contempt. In particular I knew that if I had for a moment imitated these refugees in their backstage conduct as well as in their frontstage elegance, and if my parents had got even a suspicion of it, I should have been relegated to outer darkness. Finally my parents became aware that there was something a trifle contemptuous of ourselves in the Chekhov garden party behavior of these exotics, and their visits became rarer and rarer and then ceased altogether.

My father had been desperately opposed to the Communists from the beginning. At least part of this was because his intimate connections with Russia were with such men as Milyukov, who was a Menshevik and associated with the unsuccessful Kerensky regime. The ideal thing for my father to do would have been to retain sufficient contact with the new Russia, whether he liked it or not, to enable him to understand the detail of what was happening, and even to offer warning to the Government of the United States of whatever might have been the new dangers. At any rate, from the time of the revolution on, not only did my father's researches drift further and further away from Russia, but thread after thread of his personal contact with that country was broken. He published a book on Russia and how it might be seen from the American point of view, but it was based exclusively on water that had long since run under the bridge. In short, Father's alienation from Russia led to an alienation from his own Harvard responsibility vis-à-vis Russia, and I have no doubt that his defection from Slavic researches was a factor that later led Lowell to view unsympathetically any request of my father to continue his Harvard connection well beyond the normal age of retirement.

Public opinion was turning even more toward the Allies, and it had become most likely that we should enter the war on their side. In the second term, there was founded an officers' training organization known as the Harvard Regiment, which I promptly joined. In the depth of winter, we would trudge through the snow in sleazy summer uniforms to the baseball cage on Soldiers' Field, where we were initiated into the School of the Soldier and the School of the Squad. When spring came, we continued our exercises out of doors behind the Harvard Stadium, and made several routine marches and other excursions. We also marched out to the State Rifle Range in Wakefield, where for several days we were trained in the art of musketry. Despite my poor eyesight, I made for once in my life the grade of sharpshooter. This does not redound to my credit but rather to the credit of my instructor, a Mr. Fuller, a Boston broker.

The Harvard Regiment left us hanging in the air, but I had plans to go to Plattsburg in the summer and train for a commission in the reserve. All this was, of course, dependent on my finding a job for the ensuing year. I met various deans and heads of departments looking for new staff candidates, but none of them seemed particularly interested in me, and I was assured by Professor Perry that I was not good enough to merit much of a recommendation. I was not a very promising bet at that time, but I cannot help believing that some part of my department's coolness was based on my lack of years and

TRIAL RUN

on a conservative unwillingness to experiment with the unknown.

Yet my difficulties were in part a consequence of my development. A year earlier, my assistantship had come easily, since it did not imply any competition with older and more approved Harvard men for anything they really wanted. Now, at the end of my year, matters were different. I was asking for an instructorship, and with it an entrée to a career in the field where desirable jobs were few and far between. This was more than Harvard teachers were willing to bestow on someone hard to manage and without a clear pattern for his future.

Finally, under pressure of my father, I decided to look for a job in mathematics rather than in philosophy and in a way I considered rather humiliating: by registering my name with a number of teachers' agencies. It is a procedure akin to fishing, in which the nibbles are much more frequent than the bites. Finally, I did get a bite, and I agreed that I should spend the next year as an instructor of mathematics at the University of Maine in Orono, Maine. We returned to The Top of the World, our summer home in Sandwich Township, for the summer.

We had another visit from Raphael Demos, and from Jim Mursell, a young Australian student from the Harvard philosophy department. Mursell, Demos, and I went on another tramping expedition north to Mount Washington by way of the Webster train. When that was over, I went to the Officers' Training Camp at Plattsburg, New York, to try for the commission in the Army that would be so desirable when the United States should enter the war.

I took the autostage from Sandwich Lower Corner. On the lake steamer across Lake Champlain I found a youthful school fellow, the unregenerate scoundrel from Walker Street who had chased another boy with an ax, and who had become one of the most promising young swindlers in Massachusetts. He tried to palm off on me as an officer a cavalry private with whom he happened to be traveling, but I had become sufficiently acquainted with army insignia not to be completely misled by the resemblance of the yellow hat-cord of the cavalryman with the gold-and-black hat-cord of the officer.

The Harvard Regiment had prepared me in some measure for the army camp. But I was still somewhat shocked by the bottle-drinking and foul-mouthedness of even those gentleman sham-soldiers. There were only one or two men in my company with whom it interested me to talk, although more than one New York society buck was present. The company mate who attracted me most was a member of a missionary family from Burma, who represented a continuation of the great missionary tradition of Adoniram Judson.

My mountain hikes had put me in good training, and I was in reasonably hard condition to stand the route-marches and sham battles. I was astonished to observe, even in myself, the difference in our conduct caused by the fact that we were members of a large, similarly directed group. For example, I would normally never think of bathing naked by the side of a traveled highway. However, when there are a hundred naked bodies already in the river, one cannot perceive in one's own nakedness any particular additional insult to public decency.

Again, one day when striking across the company street of pup-tents, I accidentally crushed a man's glasses. My normal instinct would have been to make myself known to him and to pay for them; but in the presence of so many uniformed and not very responsible youngsters, I am afraid that I hit and ran.

I was rather miserable during the part of our training devoted to musketry. Without the special coaching that I had received from Mr. Fuller when I was in the Harvard Regiment, my eyesight did not permit me to hit a barn out of a flock of barns. When I explained my shortcomings to the musketry officer and went back to my tent, my tentmates began to accuse me of malingering. They had already learned how easily they could make me squirm by their obscenity, and I was completely miserable. I was so angry that I laid my hand on one of the rifles stacked in the tent, with no intent whatever to use it as a rifle and very little intention of using it as a club, but more as a gesture of anger and despair than anything else. Of course, they disarmed me without any trouble, but I was unspeakably shocked when I saw clearly for the first time the murderous construction that could be put on my actions.

I finished the camp without being recommended for a commission, and with no particular feeling of accomplishment. I returned to the mountains for a week or so, and then made my way to Orono, and to my new job at the University of Maine.

I found Orono a rather raw and less inviting copy of the New England towns to which I had been accustomed. I made arrangements for board at the Orono Inn, which was a sort of commons for junior faculty members, and took lodgings in a rather attractive white New England house owned by the university librarian.

Although it was a great satisfaction for me to have a job beyond the immediate supervision of my father, I was not happy at Maine. The older, permanent professors were for the most part broken men, who had long given up any hope of intellectual accomplishment or of advancement in their career. A few of them still showed ragged traces of cultural ambitions, but the greater part were resigned to their failure. The younger men were almost all transients like myself, who had been bought wholesale at teachers' agencies after the cream of the crop had been skimmed off through the special efforts of their professors. The men who remained, the traveling guests of the university, had no interest whatever in the place, and had as their sole ambition to leave it as soon as possible, before their employment there should fix on them too definitely the stigma of unemployableness at more desirable institutions. Few and far between are the individuals who do not perish of intellectual atrophy at such places.

The president was an importation from the Middle West, and well aware of his authority. The students in those days were largely a strapping lot of young farmers and lumberjacks, who managed to be quite as idle and collegiate as the students of universities of the Ivy League, but at one-third the expense. Their sole interests were in football and in nagging the lives out of their professors. As I was young, nervous, and responsive, I was their chosen victim. Most of my courses were dull routine to them, and many is the penny which I heard dropped in class to annoy me.

Examination cribbing and the copying of homework had been reduced to a system. I soon found that to report such irregularities, as I had been told was my duty, reflected much more seriously on me than on the miscreants. I also found that some of my colleagues, within and without the department, resented my ignorance of and indifference to the very rigorous social protocol of the small college, my precocity, and what they considered my intellectual pretensions.

I tried to get back to mathematical research work. Dr. Sheffer of Harvard had recently suggested to me a way in which mathematical logic could be based on one single fundamental operation. I followed out his suggestion in a somewhat modified form, and published a paper in my name only. I think I gave credit to Dr. Sheffer in the paper, but I do not think that I did so adequately. I now see that my modification of Sheffer's work was scarcely sufficient to have merited publication as a separate paper, and that I should have waited until he had staked his claim in a more definite form. It is not only medicine and law that have a rather rigid code of ethics, and all the good will in the world will not make such a code come to one as a habit until he has at least a certain modicum of experience of the customs in the matter. Luckily, neither Dr. Sheffer nor my other colleagues in mathematics held my offense against me, but it worried me deeply when I came to realize what I had done, and it has been on my conscience to some extent to this day.

My parents were very much discontented with the generally gloomy tone of my letters from Orono. However, I must say that they put themselves out to give me a good time during my brief absences on ticket-of-leave. It was during these times that I learned the beer-and-sauerkraut pleasures of Jacob Wirth's restaurant in Boston and the delights of the new theatrical repertory company which had just been established at the Copley Theater. I also saw more of the rather primitive movies of the time than I had ever seen before, and was now and then allowed to meet some of the Radcliffe fellow students of my sister Constance. But even in the midst of my Boston amusements, there was the dread of my impending return to Orono.

Ultimately I was accepted as a member of a small research group at Orono and the neighborhood. The chief spirit of the group was the statistician, Raymond Pearl, who later had a distinguished career in the medical school of Johns Hopkins University. In his little house by the trolley line which ran between the village and the university, the chosen guests could hear good talk and a proper valuation of ideas. In those days when the English Cambridge seemed infinitely far behind me and the prospect of a civilized career infinitely far ahead, those visits to Dr. Pearl's house made me feel alive again.

Another of the few scholars of the University of Maine was Miss Boring. She was a zoologist and a sister of the psychologist Boring who had been my fellow graduate-student at Cornell. I was again to meet Miss Boring many years afterward in China, when she was teaching at Yenching University and I was teaching at the neighboring Tsing Hua University.

There were also a few doctors of the Bangor General Hospital in our group. I remember some very interesting lectures on pulmonary cancer, which were given long before this mimic of tuberculosis had been generally recognized as a clinical entity.

The meetings of this group did not constitute the only occasion for my going down to Bangor on the trolley. Bangor was no longer the riproaring town where the returning lumberjacks first tasted the pleasures of women and bootleg liquor, but the bad old days had put a blight on it and it lacked the charm of more favored New England cities. What called me there was a military drill corps whose meetings in a gymnasium were attended by good portly Bangor citizens much older than I.

The winter trolley ride to Bangor and the train trips to the city had a charm of their own. The whole landscape was blanketed with deep white snow, which was so cold that it creaked when you stepped on it. The air shocked the lungs like a cold fire. The time had not yet come when the auto roads should be kept open all the winter, and the sound of sleighbells was in the thin air.

There was a Norwegian student living in the same lodging house who was specializing in paper manufacture. The University of Maine was a center of instruction in this subject, and indeed the air reeked of the sulfurous fumes of the local paper mill. The Penobscot backwater opposite our house was divided by the cribs and booms which held back a supply of pulpwood floated down from the north. My Norwegian friend used to ski to school, and to make many ski trips across the snow-filled marshes and woods. The rest of us, students and professors alike, had still to learn this Nordic sport, and went to our classes on the snowshoes manufactured by the Indians in Oldtown nearby. There was always a forest of snowshoes stuck tail down in the snowbanks outside the doors of the college buildings, and the students, boys and girls alike, wore to their classes the woollen stockings and the Barker boots or pac-moccasins which mark the northern woodsman.

Time hung heavy on our hands. I remember reading through the whole of the writings of O. Henry and Mark Twain in a dark corner of the university library stacks, but unfortunately the present vogue of the detective story had not yet developed. I was greatly distressed during the winter both by the coming participation of America in the war and by the news of the death of my friend Everett King. He had been my companion in much boyish experimentation, and I am certain that if he had lived he would have become an important figure in American science.

Spring began to break with the wet suddenness which is characteristic of spring in northern New England. There were one or two new faces to be seen about the campus. I remember one pleasant newly arrived young American married to a French wife. He endeared himself to me by letting me accompany him on fishing excursions to the edge of the vast wilderness on the other side of the Penobscot.

There was little question that we were soon to enter the war. The existing officers' training corps was much enlarged, and every possible drillmaster was pressed into service. In view of my experience in the Harvard Regiment and at Plattsburg, I too was pressed into service; but I proved to be both too unskilled and too peremptory in my commands, and I did not make much of a success of it. When the war finally came, I asked to be released from my teaching responsibilities to enter some branch of the service, for I was no less eager to leave Maine than the university was to see the last of me.

I passed some sort of preliminary physical examination with the friendly connivance of a Bangor doctor, and left on the steamer for Boston to try my fortunes with the services.

On the way it occurred to me almost for the first time that I might be taking a real risk both of life and of limb, and I was distressed. However, I reasoned with myself that I had a good chance of coming out of the war with some sort of usable body attached to my soul.

When I arrived in Boston, I made the rounds of the harbor forts and the enlistment bureaus, in the hope that I could enter one of the services, if not as an officer, at least as an enlisted man. My eyes were against me everywhere. Finally my parents decided, with my acquiescence, that I should try for an army commission in the R.O.T.C. which had just been established officially at Harvard.

With the coming of the war, the new Officers' Training Corps became a much more systematic outfit than the old Harvard Regiment. We were quartered in President Lowell's new freshman dormitories, which were later to become the units of the Harvard house system.

We received some special lectures from a group of academically trained French Army officers, one of whom, Major Morize, remained for many years as professor of French at Harvard. During the summer, we entrained for Barre Plains, where we encamped and held maneuvers. I remember some of our trench-digging, mock fighting, and bayonet instruction. I was there only part of the time because examinations for commissions in the artillery were being held in the new buildings of the Massachusetts Institute of Technology. I knew that this was about my last chance to obtain an artillery commission for the war. Naturally, I did rather well in my mathematical examinations, but I could not show any particular military aptitude. I failed dismally in my physical examinations and in an examination on horseback riding at the state armory. I came totally unprepared for this, and I fell off an old nag which was as steady as a gymnasium horse.

TRIAL RUN

As to my physical examination, my eyes would have damned me anyhow, but I also showed a blood pressure which was high for my age, although my continued survival proves to my satisfaction that it was not within any dangerous limits. The Army doctors probably took it quite correctly as showing that I was of too labile a temperament to be good army material. What chances I might have had for a commission were damned by the fact that, in accordance with the pseudonoble mores of the time, I tried to bluff and argue one of the doctors into giving me a favorable medical report, so that he frog-marched me ignominiously out of the room.

I graduated from the R.O.T.C. with a document that was eminently not negotiable for a commission. The end of the summer was near, and I spent what was left of it at Silver Lake in New Hampshire. I did some reading on the algebraic theory of numbers, which I had begun to study while I was in Maine, and I made a few attempts to extend Birkhoff's results on the four-color problem.

This, together with Fermat's last theorem and the demonstration of Riemann's hypothesis concerning the Zeta function, is one of the perennial puzzlers of mathematics. Every mathematician who is worth his salt has broken a lance on at least one of them. I have tried to solve all three, and each time my supposed proof has crumbled into fool's gold in my hands. I do not regret my attempts, for it is only by trying problems that exceed his powers that the mathematician can ever learn to use these powers to their full extent.

We lived near Professor Osgood's summer home that summer, and I saw a great deal of him. He was a much more genial personality in a New Hampshire summer cottage than he had seemed to be in the full glory of his Harvard position. I also did a bit of mountain climbing, and it is the proudest exploit of my youth that together with my sister Constance and a friend, I covered thirty-four miles in one day in a tramp over Mt. Passaconaway, Whiteface Mountain, and back. Of course, I was exhausted and feverish the next day; but one is resilient in one's twenties, and the effort seemed to do us no harm.

« XVII »
MONKEY WRENCH, PASTE POT, AND THE SLIDE RULE WAR

1917–1919

WHEN WE RETURNED to the city, it was obvious that I should have to look for some civilian form of war work. My search for nonacademic employment was precipitated by the war and by the nearly total suspension of normal university life which it involved. I felt that my mathematical training was the most useful thing I had to offer, so I took the trolley out to the Fall River Shipyard in Quincy to see if I might not learn to be of use in ship-propeller design. Nothing came of this, so I made a similar trip to the General Electric factory in Lynn. One of the engineers there was a Russian friend of my father, and another had been one of the instructors in my physics course at Harvard. I naturally met with a more favorable reception. I was told that I could be of no use immediately; but if I were willing, they would take me on as a paid apprentice in their program of engineering training. This meant that I took on the moral responsibility of staying for two years. I accepted, and started to work in the turbine department. I helped run some steam-consumption tests, and used a little mathematics in some thermodynamic problems. When I returned home each day, I was tired but happy, and hopelessly begrimed by that grease which belongs to an engineering factory and which no soap seems able to remove. I took this as the badge of the honest workman.

However, my father was convinced that with my clumsiness I could never really make good at engineering, and began to look around for other work for me. He had written an article or two for the *Encyclopedia Americana*, then located in Albany. He secured from Mr. Rhines, the managing editor, an invitation for me to join them as a staff writer. Though I felt morally obligated to stay on at the General Electric Company, I was too dependent on my father to dare to contravene his orders, so I had to present my shamefaced resignation to the engineers who had given me my chance in Lynn. I was told that I never could come back to a job there, but with my helplessness and lack of independent experience, there was precisely nothing that I could do about it but to follow orders.

Father accompanied me to Albany and saw me placed with a rather pleasant landlady in an old high brick house not far from the State House. He also accompanied me to Mr. Rhines's office. This was situated several flights up a freight elevator in a dismal business building facing on a cabbage-strewn market of the European type. Mr. Rhines was a bearded old gentleman, efficient and strict but kindly.

Albany appealed to me from the beginning. In many ways the downtown part resembled a European city, or Boston's Back Bay. I found good restaurants at which to eat, a good vaudeville theater, and good movies at which to pass the time. I also found a gymnasium at the local Y.M.C.A. at which I could keep myself fit.

My *Encyclopedia* work consisted in the compilation of some of the less important short articles; and I believe that it was paid on a piece basis. I soon found that I was working in a not

MONKEY WRENCH, PASTE POT, AND THE SLIDE RULE WAR

too uncomfortable Grub Street, with a group of cheaply paid colleagues who were either on their way up or on their way down. Among us was an elderly English businessman who had failed once and was now too old to start on a new business career. He prided himself on his knowledge of the Gilbert and Sullivan operas and his ability to write both words and music in the same style. Another of our writers had risen from engine-driver on a British railway to become librarian (or keeper of the morgue) for the London *Times*. He had a portfolio full of proof sheets of obituary notices of famous men still alive, written at the time of their illness, or (if they were sufficiently important) permanent documents in case of their sudden death. He had lost his position through drink, but had still enough ability left to make himself very useful on the *Encyclopedia*. His fund of stories was mostly improper, but generally improper in an amusing way.

We had also an Irish ex-seminarian, belonging to the Dublin of James Joyce, who sported that delightful literary English of the educated Dubliner, with just the hint of the possibility of a brogue. There was a young American lexicographer who used to play tennis with me and later became chief editor of the *Encyclopedia*.

There was a young girl graduate of Cornell among us. She was the daughter of a Russian-Jewish Albany fur merchant. I found her very attractive and intellectually stimulating. We used to take walks in the country about Albany, and I would call on her at home and take her to the visiting theatrical shows. She and I were two members of the group young enough to be definitely on the up-grade. We enjoyed the eccentricities and the friendliness of the older people who had found a refuge in this queer but agreeable *faubourg* of Bohemia. Nevertheless, as approximate contemporaries there were many minor amusements of the town which we shared with one another rather than with others of the *Encyclopedia* group. Even when

I found that she was engaged to a young doctor who was in the Army in France, I continued to take her out from sheer hunger for female companionship.

We used to alternate our work between the office and the New York State Library in the Educational Building near the Capitol. This building held a real fascination for me. Besides being a really good library, and, I believe, the headquarters of the State Board of Regents, it contained the New York State Museum with its mixture of geographical, geological, anthropological, paleontological, botanical, zoological, and petrographical-mineralogical-crystallographic collections. I used to spend much of my spare time in the museum, and perhaps a good deal of time that should not have been so spared. I made the acquaintance of one of the curators, who was a specialist in crystallography and in gems, and burrowed into the article on crystallography in my beloved *Encyclopaedia Britannica*. I also saw a good deal of one of the state paleontologists. These contacts rearoused my interest in the origin of the vertebrates, and I reread Gaskell and Patton to see if there was any possible way of making sense of the arachnid theory of vertebrate origins.

I found that the hack work of compiling an encyclopedia has a peculiar ethics of its own. A compiler should be exactly what the word implies. It is permissible to use information from other existing encyclopedias, if it is carefully collated with outside sources. If you must crib, crib from foreign-language encyclopedias, and do not favor any single source excessively. Be very careful before you give any article a by-line. In general, eschew your own original ideas.

It took me some time to become so familiar with the rules that they were at last instinctive; and more than once I succumbed to the boyish urge to cut corners. I reinforced my habit of reading encyclopedias by the habit of writing them, and the considerable scope of my existing information was

MONKEY WRENCH, PASTE POT, AND THE SLIDE RULE WAR

useful. In one or two articles, such as a part of the one on "Aesthetics," I deviated into individual philosophical writing, and the material still looks fresh to me at the present day. I toyed with the idea of collecting into a small book a number of such articles and of my previous writings on philosophical topics. Besides such reasonably original and good articles, however, I ventured to write on a number of subjects on which I was not adequately prepared. Some of the articles I submitted on ballistics were the most utter balderdash. I hope that Mr. Rhines did not let them get through.

With all the shortcomings and unpleasant sides of hack writing, it was a wonderful training for me. I learned to write quickly, accurately, and with a minimum of effort, on any subject of which I had a modicum of knowledge. In revising my material, I learned the marks and the technique of the proofreader.

The problems of literary style are curiously related to the problems of speaking a foreign language. The experience which one acquires in hack work is quite parallel to the experience of being immersed in a foreign milieu and having to speak the language day after day. For it cannot be denied that literary English as it should be written, although it has deep roots in colloquial English and can only depart too far from the colloquial at its peril, is in many ways a distinct language. For example, the richness of metaphor and of term of speech appropriate to an effective literary style would be heavy and "donnish" in spoken English. Thus the problem of writing for a man who already can speak with reasonable effectiveness is to develop the same freedom in this sophisticated milieu that he already has in his daily speech. If I am to speak Spanish effectively, I must think in Spanish and I cannot be tempted to translate out of an English phrase book. I must say the sort of things that a Spanish-speaking person would say, and these are never precisely the same as those which an English-speaking person would say. Similarly, if I am to write a poem or a novel or a scholarly essay with any ease, I must have had sufficient practice in expressing myself in the suitable form of the language so that my written or dictated words come out in what is already effective poetic English or the effective language of the novelist or that of the philosophical essayist. My metaphors and terms of speech should find themselves without having been looked for, not indeed in their final polished form but in a form not too remotely different from that. I do not deny the virtues of revision and of the progressive weeding out of weaknesses and faulty expressions. Nor indeed do I wish to prescribe to other writers what is of necessity an intensely individual expression of their own thinking. But at least as far as I am concerned, whether I work in mathematics or in writing, I cannot arrive at the fullest expression of my own consciousness until I have already penetrated considerably below the level of consciousness.

I was happy in Albany. I liked my co-workers and employers, I liked the work, and I liked my new sense of independence. In view of the difference in the nature of my new work from that of my father, I was even less under parental pressure and criticism than I had been at the University of Maine. I was also older and more fit to stand on my own feet. Compared with Orono or Bangor, Albany was a paradise of neatness, tradition, and civilization.

Behind my new-found happiness and content, there was always the hollow echo of the war. The R.O.T.C. experience had proved to me my essential unfitness for a commission, but I had still some hopes of being admitted for limited service as a private under the new draft. For the time being, I joined the New York State Guard. This was the organization that occupied the armories when the National Guard went out to fight, and it had the duty of patrolling water supplies and powerhouses. I was not called on for such semiactive duty, but

MONKEY WRENCH, PASTE POT, AND THE SLIDE RULE WAR

my earlier training in drill stood me in good stead inside the armory. When spring came, we used to spend Saturdays and sometimes even Sundays on an island in the Hudson where there was a rifle range.

I used to return to Cambridge for short vacations from time to time. My sister Constance made some effort to take me into the social life of her Radcliffe friends, and I remember that there was an Australian girl with whom I went out occasionally. I spent a short summer vacation in New Hampshire, then returned to my work on the *Encyclopedia*. But it was by now apparent that this sort of work, while very acceptable as an instructive effort and a way station on my career, would be very unsatisfactory as a terminus.

I was not distressed to find myself engaged in work that was enjoyable even though it led to a dead end. Such work was in many ways in conformity with my actual age. From the standpoint of the dramatic movement of my life, this may seem to be a retrogression but, in fact, I do not think that it was.

In the life of the individual, neither the standard success story of the slicks nor the Greek tragedy is the normal and the significant outcome. That the individual dies in the end is clear, but what is also clear is that this fact of the physical termination of one's life is not its significant outcome. In the course of the voyage from the nothingness before conception to the nothingness of death lies everything that is really important in life, and this voyage is generally neither a dramatic steering into the maelstrom nor a triumphant ascent from success to success, without some intervening setbacks and some periods of calm cruising.

It may seem a step down to follow years of precocity and early academic degrees by the somewhat routine tasks of a shopworker, a hackwriter, a computer, and a journalist. Yet these were for me the equivalents of experiences in the world about me which many a boy obtains much younger as a prosaic part of his normal development. Just because I did not have these experiences at an earlier period, and just because this contact with the world about him is essentially part of the education of everyone, these everyday experiences had for me a glamour and a novelty which they would not have had for a boy of more usual bringing up. Thus a description of these periods, and of their reaction on me, is as essential to this book as all its other and perhaps more exciting parts.

When I started to look for a job again, I heard of a vacancy in mathematics at the University of Puerto Rico. I sent in my name as a candidate, but I did not get even a nibble at my trial balloon. A few days later, I received an urgent telegram from Professor Oswald Veblen at the new Proving Ground at Aberdeen, Maryland. He asked me to join their ballistic staff as a civilian. This was my chance to do real war work. The demand was immediate, so I saw Mr. Rhines and terminated my connection with the *Encyclopedia*. I took the next train to New York, where I changed for Aberdeen.

Aberdeen, Maryland, was then a minor country town of no particular distinction. A little branch railroad line, operated by the Government, ran from the village to a place on the line as yet without a railroad station, where I got off. I found a collection of temporary wooden huts and quagmire streets, which has since become a very pretty little Government station. In those days a tractor was always kept ready to pull the squelching trucks out of the streets.

The establishment of Aberdeen Proving Ground constitutes a very important epoch, both in the scientific history of the United States and in the private careers of the scientists who were stationed there. Although American scientific work had long been important in astronomy, geology, chemistry, and some other fields, most of our best men had been trained in

MONKEY WRENCH, PASTE POT, AND THE SLIDE RULE WAR

Europe, or even imported from Europe. Mathematics was well behind these other sciences in its American recognition. As I have said before, Birkhoff was the first really great American mathematician to come to the top without European training. It was now only six years since he had taken his degree in 1912. Thus we American mathematicians were a feeble race, and the country at large regarded us as so many useless fumblers with symbols. It was scarcely possible to believe that we could play any useful part in the national war effort.

The war with Germany involved the design of many new types of artillery and of ammunition. For each type of artillery, and each type of ammunition used in it, it was necessary to construct a complete new range table, and to put it in the hands of the men in the field. These range tables consisted of lists of the ranges to be expected from the gun and the ammunition for each angle of elevation, together with corrections for tilt of the trunnions, unit change in the angle of elevation, for unit excess of powder charge or ammunition weight, for wind, for air pressure aloft, and so on. The tables also contained estimates of the probable errors of all the primary data. The old-fashioned methods of computing range tables had proved both too slow and too inaccurate for modern needs, and had broken down completely in the new and very exacting field of antiaircraft fire. Thus there was urgent need for every available man with mathematical training to operate a computing machine, and civilians like myself were pressed into service, drafted mathematicians were transferred to the Ordnance Corps and to Aberdeen, and even officers were recalled from the front to sit at a desk and to work a slide rule.

Professor Oswald Veblen of Princeton was made a major in the Ordnance Corps and put in charge of this motley group. His right-hand men were Captain F. W. Loomis and First Lieutenant Philip Alger, later Captain, whose father had been the great ballistic expert of the Navy. As for the rest of us, we lived in a queer sort of environment, where office rank, army rank, and academic rank all played a role, and a lieutenant might address a private under him as "Doctor," or take orders from a sergeant.

One thing is clear, however: we accomplished the task expected of us. It was a period in which all the armies of the world were making the transition between the rough old formal ballistics to the point-by-point solution of differential equations, and we Americans were behind neither our enemies nor our allies. In fact, in the matter of interpolation and the computation of the corrections of the primary ballistics tables, Professor Bliss of Chicago made a brilliant use of the new theory of functionals. Thus the public became aware for the first time that we mathematicians had a function to perform in the world. We still did not count in its eyes as magicians comparable with the chemist and engineer.

We were fortunate in this respect, for while our newly won prestige contributed appreciably to our salaries and our ease of finding employment, the authorities did not yet consider us sufficiently important to be worth the trouble of interference and annexation. Emerson did not tell the whole truth about the fate of the man who devises a better mousetrap. Not only does the public beat a path to his door, but one day there arrives in his desecrated front yard a prosperous representative of Mousetraps, Inc., who buys him out for a sum that enables him to retire from the mousetrap business, and then proceeds to put on the market a standardized mousetrap, perhaps embodying some of the inventor's improvements, but in the cheapest and most perfunctory form which the public will swallow. Again, the individual and often delicious product of the old small cheese factory is now sold to the great cheese manufacturers, who proceed to grind it up with the products of a hundred other factories into an unpleasant sort of vulcanized protein plastic.

MONKEY WRENCH, PASTE POT, AND THE SLIDE RULE WAR

In the Second World War and the days that have followed it, the very success of the American scientist has driven him the way of American cheese. This war saw every chemist, every physicist, every mathematician, dragooned into Government service, where he had to put on the blinders attendant on working on classified material, and to confine his efforts to some minute sector of a problem of whose larger implications he was held deliberately ignorant. Although the excuse given for this procedure was that it kept secrets out of the hands of the enemy, and this was undoubtedly part of its real intention, it was not and is not entirely unconnected with the American lust for standardization, and the distrust of the individual of superior abilities. This, in turn, is related to our love for Government projects or private laboratories with budgets running into the millions of dollars which put a premium on the traditional Edisonian search through all possible materials, at the expense of erratic and unpredictable use of reason and intellect.

However, in the early days of Aberdeen Proving Ground, the King Log of indifference was already dead, and the King Stork of regimentation had not yet ascended to the throne. It was a period of nascent energy in American mathematics. For many years after the First World War, the overwhelming majority of significant American mathematicians was to be found among those who had gone through the discipline of the Proving Ground. I am speaking of such names as Veblen, Bliss, Gronwall, Alexander, Ritt, and Bennet.

It was the young men who particularly interested me. I found Hubert Bray there, whom I had last seen when I was graduated from Tufts College. Bray has been associated with Rice Institute for many years and is now head of the mathematics department there. For a while we lodged together. Later I shared a roughly boarded compartment in the civilian barracks with Phillip Franklin, now my brother-in-law and colleague at the Massachusetts Institute of Technology, and with Gill of the College of the City of New York. I also had as companions for a shorter period Poritzky, who later gave up pure mathematics and academic work for applied mathematics and a job with the General Electric Company, and Widder, now of the Harvard mathematics department.

This list is far from complete. Graustein, who had left Harvard for the Proving Ground and finally left the Proving Ground for an officer's commission, was a leading mathematician at Harvard for many years before his untimely death. I have also left out the names of a large number of astronomers, engineers, and secondary school teachers, with whom I have come rather less in contact in recent years.

Franklin and Gill, nineteen and thus considerably younger than I, were my particular cronies. When we were not working on the noisy hand-computing machines which we knew as "crashers," we were playing bridge together after hours using the same computing machines to record our scores. Sometimes we played chess or a newly devised three-handed variant of it on a board made of a piece of jump-screen, or risked the dangers of burning smokeless powder or TNT. We went swimming together in the tepid, brackish waters of Chesapeake Bay, or took walks in the woods amid a flora that was too southern to be familiar to us. I remember the pawpaws and their exotic tropical manner of growing their fruits directly from the tree trunk.

Whatever we did, we always talked mathematics. Much of our talk led to no immediate research. I remember some half-baked ideas about the geometry of Pfaffians, in which I had become interested through Gabriel Marcus Green of Harvard. I cannot remember all the other subjects we discussed, but I am sure that this opportunity to live for a protracted period with mathematics and mathematicians greatly contributed to the devotion of all of us to our science. Curiously enough, it

MONKEY WRENCH, PASTE POT, AND THE SLIDE RULE WAR

furnished a certain equivalent to that cloistered but enthusiastic intellectual life which I had previously experienced at the English Cambridge, but at no American university.

I made several trips home on furlough. I saw a lot of G. M. Green on these trips. He had become very much devoted to my sister Constance, who had become a budding mathematician herself. On one of these trips I talked over with my parents a plan which had long been in mind, which consisted of using my contact with the Proving Ground as a means for enlisting or being drafted into the Army for limited service. Finally, in October of 1918, the opportunity came, and with Major Veblen's co-operation I went to a neighboring market town and county seat to receive my enlistment papers.

I was sent to the recruit depot at Fort Slocum on an island off the coast of Westchester County, New York. It had by now become clear that the war was drawing to an end. I was appalled by the irretrievability of the step I had taken. I felt as if I had been sentenced to the penitentiary. The crowding of the recruits and their incompatible simultaneous attitudes of scared boys and of tough swaggering young soldiers were far from congenial to me. The one alleviation of my life on the island was the presence of another equally unmilitary recruit, Dr. Harry Wolfson of the Harvard department of Semitic languages. My uniform was strained by my corpulence, but Wolfson's reached nearly twice around him. Not even these uniforms could conceal the college professor in us as we walked around the seawall, discussing Aristotle and medieval Jewish and Arabic philosophy.

At length a group of us were shipped back to Aberdeen Proving Ground. We rode around Manhattan in a tugboat and entrained for Philadelphia from one of the stations on the Jersey shore. In Philadelphia we heard the blowing of steam whistles to celebrate the first false reports of the armistice with Germany and saw the showers of paper floating down from office windows. About two days later, when we were already assigned to our companies at the Proving Ground, we were assembled early in the morning and told that the true armistice had been signed.

The military setup at the Proving Ground was most peculiar. Besides the administrative groups, the ballistic group, and a few other technical groups of the same sort, there was the powder-bag-sewing group and a large group of laborers to do the necessary digging and construction work. The latter were largely men who were not sent to the front because of venereal disease. All the different groups were mixed together indifferently in the companies and in the barracks. I do not need to expatiate on the continual shock of the impact of this foul-mouthed crew on a man who was not yet fully adjusted to the brutal frankness of army life.

I did two turns of guard duty. I got off easily on one of them, carrying a watchman's clock throughout the night in the building which housed the chronoscopes and the scientific library. Between my rounds I found plenty of interesting material to read. The other time I was an ordinary sentinel out of doors, carrying a rifle with a fixed bayonet. I found it hard not to drowse off and to keep sufficiently alert to challenge the Officer of the Day. In the small hours of the morning I had a little rest on the bare bedsprings of a cot in the guard room; and although I felt like a waffle when I woke up, the relief of the big cup of steaming coffee which was passed around with some cheese sandwiches more than made up for it.

Besides such military duties and my work in the office, I did a certain amount of work at the firing "front," collecting range-data for antiaircraft fire. We had a special telephone line connecting the gun station and two or three observing stations, where observers looked through peep sights at the reflections of the shell bursts in flat horizontal mirrors covered by co-ordinate gratings. In view of my inferior eyesight, I was the gun telephone operator; and I lay upon an earth bank, unpleasantly near to the noise and the blast of the gun-muzzle, and gave the observers the time of the firing of the gun, the burst of the shell, and five-second intervals thereafter. These later intervals enabled the men at the mirrors to synchronize their observations of the drift of the smoke downwind, and thus to compute the wind aloft. I also notified the gun crew when the observers were ready.

The observers rode out to their stations in an old Ford beach wagon, and sometimes they had to cross a range where firing was going on. Theoretically, the safety officer should have interrupted the firing to let them go by; but even safety officers grow careless with time, and this precaution was not always observed. I remember one time when the observers at a tower down the range complained that shrapnel was coming through the roof. "All right," said the safety officer, "we'll just fire another couple of rounds and call it quits."

As long as we had felt that we were doing work necessary to win the war, our morale had remained high. After the armistice, we all felt that we were marking time; and in particular those of us who had enlisted at the last moment considered ourselves a pack of fools. The civilians began to leave at the first opportunity, while the rest of us were going through the motions of military service until we could be sent to a camp where we should be discharged from the Army.

Even with a temperament not suited for a regimented life and a more than average desire to see what I was doing and to know what it meant, I had found a few months of army life a haven at a time when I had been very tired for years from making my own decisions. It has been said many times that the motives of the soldier and the monk are curiously similar. The love of a regimented life and the fear of personal choice and responsibility make some men feel secure in the uniform as well as in the monk's hood.

I was very deeply curious as to how the war was to end and what the new postwar life was to be. Meanwhile, I marked time. This emotional stagnation affected me both before my induction into the service when I was already leading much of the life of an army camp, and afterward; but its main impact began to fade out with the temporary letdown of the armistice and the hope that things could get back into their civilian channels.

While waiting for my orders to Camp Devens at Ayer, Massachusetts, for my final discharge, the influenza epidemic hit us. At the beginning we did not take it seriously, but soon it became a commonplace to ask about a soldier and to find that he had died the day before. We all wore flu masks, and the elephantine Professor Haskins of Dartmouth smoked his pipe through his. One very honest and conscientious soldier, an M.I.T. graduate who had been put to work unloading a freight car, complained that he was unwell. The doctor sent him back to his work, and he died of pneumonia the next day.

It was a gloomy experience to see the rough pine coffins piled high on the station platform, and to wonder where the next blow would fall. I received a telegram from my father, telling me that my friend Dr. G. M. Green of the Harvard mathematics department, and betrothed to my sister Constance, had just died as the result of the epidemic. This news hit me hard. It came just before I left for Camp Devens.

I found Ayer superficially the town of my youth, but much changed in every important way. With the tendency of the railroad to lengthen the locomotive runs and the blight that had already begun to fall on the branch lines, Ayer was no longer the important railroad center it had once been. On the other hand, Camp Devens, which was a newcomer since the beginning of the war, was considerably larger than the town itself, and the merchants fattened on the soldier trade.

There was not much to do while waiting for one's discharge.

MONKEY WRENCH, PASTE POT, AND THE SLIDE RULE WAR

There were medical examinations to take and papers to sign. I put in one day working on the coal pile of the power station. I spent much of my time at the various post libraries, reading the writings of G. K. Chesterton. At length the day of my release came, and after a brief visit to my friends at Brown's drugstore, I took the train home.

« XVIII »

THE RETURN TO MATHEMATICS

THE END of the war brought a keen sense of loss to all my family in G. M. Green's death. Green had been a charming and unassuming young man, deeply attached to my sister Constance, and of great sincerity and tenderness of disposition. He was a very severe loss to modern science, too, as he had developed a peculiarly individual vein of work in geometry, and it seemed that he was likely to furnish an element that was otherwise missing in the Harvard mathematical scheme of affairs. The death of a young man at the peak of his career is perhaps the greatest tragedy there can be, and it was hard for my sister, my parents, and me to realize that our good friend was gone.

Green's parents gave his mathematical books to Constance as the one who had shared most in that phase of his life and the one to whom they could do the most good. Constance went to Chicago in the hope of forgetting her bereavement in new work, as far as that might be possible. Hence I had the chance

THE RETURN TO MATHEMATICS

to look at these books and to read them over. They came to me at just the right time in my career.

For the first time, I began to have a really good understanding of modern mathematics. The books included Volterra's *Théorie des equations integrales*, a similarly entitled book by Fréchet, another book by Fréchet on the theory of functionals, Osgood's *Funktionentheorie*, Lebesgue's book on the theory of integration (to which I devoted particular attention), and I believe a German book on the theory of integral equations as well.

However, reading mathematical books would not pay my parents for my room and board, nor would it of itself advance me in my career. Again I began to look for a job. I sent my name to various teachers' agencies, but it was only February and no academic job was likely to start until the next September. At Silver Lake, Mr. O'Brien of the Boston *Herald* had been a summer neighbor of my parents, and they sent me to him to see if he could find a place for me on his paper. My encyclopedia experience spoke in my behalf, though I could not see eye-to-eye with Mr. O'Brien's intention to make use of my mathematical ability by training me for a job as financial editor. For this, of course, I had neither the gift nor the inclination.

The confusion between the craft of the accountant and that of the mathematician is very common. The difference between imaginative mathematics and accounting is obvious, but it is equally real between accounting and computing. The accountant works to the last cent. His duty is to eliminate discrepancies which might offer someone a chance to embezzle undetected. The mathematician works to a certain number of decimal places of accuracy. His maximum permissible error is not an absolute quantity, such as a cent, but a stated fraction of the smallest quantity with which he deals. The computer who turns accountant is likely to leave appreciable sums of money unaccounted for, whereas the accountant turned computer often works to just two places beyond the decimal point, where the logic of the problem may call for four or five, or in another case, for none. Unless a man is reasonably young and flexible it is disastrous to shift from one of these seemingly so similar tasks to the other.

Luckily, I escaped from this assignment and was put instead to straight journalistic tasks. In the Boston *Herald*, I was a feature writer, and began to make myself acquainted with the paper litter and the printer's ink, the noise of typewriters and linotypes, and the general sense of hurry and bustle which constitutes the normal background of the city newspaper. I tried writing a few editorials. I learned very soon the extreme care which the editorial writer must take to verify his facts, and to avoid unnecessary treading on toes. Then I was put on Sunday feature writing.

The textile industry of Lawrence was passing through one of its periodic strikes, and I was sent down with carte blanche to cover the larger features of the situation. I was less liberal then than I have become since. Happening to meet one of the senior Lawrence labor union leaders on the train, I was almost prepared to find that he had horns and hooves. On the contrary, he was a fine, sympathetic old Lancashire man, who had left England when the shadow of the industrial revolution was still at its darkest. He had seen the philanthropy of the early New England manufacturers give way to absentee ownership and the English weavers of the early days supplanted by French Canadians, Belgians, Italians, and Greeks. He had continued to exert his authority over the younger generation, although he had found that they needed rather different handling than their predecessors; and they had begun to develop their own labor leaders under his tutelage.

He told me to keep an eye on the living conditions in Lawrence, and on the way in which the Americanization work was handled. He also gave me a list of priests and labor union leaders in order to help me feel the pulse of the various foreign elements in Lawrence. I appreciated the sturdy honesty of the man and found his advice sound and helpful.

Lawrence was a sick town. The mills were already suffering from an overcapitalization in obsolete equipment and from the competition of the South, as yet uninhibited by any standards of employment, and favored by the lower wages made possible by a less severe climate. The owners of the Lawrence mills often had never set foot in Lawrence, and had left the problems of management and employment to their badgered agents, ground between the employers' demands for profits and the workers' demands for higher wages and better working conditions. The housing conditions were atrocious, and although the employers defended them by the hoary argument that the class of labor they employed would rapidly ruin better quarters, one might as readily reply that the foul quarters made it impossible to employ a better class of labor. I took part in one of the Americanization classes at the Y.M.C.A. and was appalled by what I saw. Not only were the teachers utterly ignorant of the language of the men they taught, which happened to be Italian in the class which I attended, but they were completely out of touch with the educated elements in the immigrant communities. The contents of the textbook in use exhorted the workers to love and to honor the boss, and to obey the foreman as if he were Jehovah himself. It was the sort of humiliating tripe which was bound to alienate any workman with a trace of character and independence.

I published my reports of things as I saw them, and they excited some opposition but, on the whole, less than I had expected. I should like to think that my articles had some influence on public opinion, making it more conscious of the importance of the housing question, and that they may thus have had some small effect on the later establishment in the

Lawrence region of garden cities, such as Shawsheen Village.

After this job, O'Brien put me on a political undertaking which was very much nearer his heart. This was to build up General Edwards, former commander of the Yankee Division, as a possible candidate for the Presidency of the United States. As far as I could see after meeting the man, Edwards was an amiable enough old gentleman of no particular distinction. I found that I did not relish my assignment at all. I visited in due course his friends and relatives in Cleveland, Ohio, and at Niagara Falls, and I looked up ex-President Taft and other notables in Washington who had known him in the Philippines.

With all my experience of hack writing on the encyclopedia, I had not learned to write with enthusiasm of a cause in which I did not believe. I was fired from the *Herald,* and the Edwards series was put in the hands of a more responsible wheel horse. I was not unready to leave the paper, but I am thankful for the experience in writing which it gave me as well as the knowledge of the American scene.

I came away from my second experience of writing for my living with a new sense of the dignity of hack writing. In general, our courses in English at our colleges are as far from teaching us to write English as our courses in foreign languages are from giving us a real mastery of any foreign language. This applies primarily to the introductory courses. They do not in general make sufficient demands on the student to bring him to the stage at which he has to write a critically acceptable thousand words every day or be punished for it severely by going without eating. They introduce him to the English language as one might be introduced to a delightful young lady at a cocktail party: the student has not quite caught her name and would not know her again. After I left the *Encyclopedia,* and especially after I left the *Herald,* I was reasonably confident that if it should ever be my duty to say anything in print, I could say it fairly correctly and forcefully the first time I should write it down.

Thus, I am glad that I had an easy test of Grub Street between my years of wandering and my years as master in my own shop. In addition to this benefit, my experience as a journeyman gave me the satisfaction of independence which I could achieve in no other way. Not only was I earning my own living, but I was doing it in a fashion that made no demands whatever on my father; and to a large extent, I was doing it away from home and parental tutelage. In short, I was growing up.

I had two mathematical papers under way to which I devoted my temporary unemployment. They both concerned the extension to ordinary algebra of Sheffer's idea of a set of postulates with a single fundamental operation. I wrote them in the stacks of the Harvard Library, near my father's office. They were published in the following year. Although they represent directions of work which as far as I know have not been followed up, they were by far the best pieces of mathematical work that I had yet written. However, I soon afterward gave up algebra and postulate-theory for analysis, which seemed to me to have a richer and firmer intellectual body. Thus it is hard for me to evaluate these papers at the present time, or even to remember exactly what they contained.

I had been trying for some years to find a publisher for my Harvard Docent Lectures. While they were in no sense a finished piece of work, I do not think that I am completely unfair in seeing in the idea of constructive logic which I there developed a certain approach to the ideas by which Gödel was able to demonstrate that in every system of logical postulates, there are theorems whose truth or falsity cannot be decided on the basis of those postulates. I finally sent the manuscript to P. E. B. Jourdain, a remarkable English logician who lived near Cambridge, and with whom I had already been in correspondence. While I was at Cambridge I had asked if I could meet him in his house, but we had not been able to make arrangements. I had not realized at the time, nor indeed until well after I had sent him my manuscript, that he was a hopeless cripple, scarcely able to move a finger. He had known very well that he had suffered from Friedrich's ataxia, a congenital nervous disorder which always ends in paralysis and early death. Notwithstanding this fatal ailment, he had married and had become editor of that important philosophical magazine, *The Monist*. He had written a very humorous and critical book concerning Bertrand Russell's philosophy, in which each chapter was introduced by an appropriate text from Lewis Carroll.

My manuscript reached Jourdain only a few months before his death. If I had known how ill he was, I would of course not have embarrassed Jourdain by sending him my manuscript. However, it was published in the periodical, and I had *The Monist* bind together in book form the reprints of my article, which had run in three numbers. It pleased me then to imagine to myself that this constituted the publication of a book.

The articles found a limited echo. Professor Broad of Cambridge referred to them; but in those days, work in mathematical logic was equally unhelpful in getting jobs in either mathematics or philosophy. Today mathematical logic constitutes a recognized career. As in some other fields, it is a career for *epigonoi*, not for pioneers. There are tables where guests are served only after the silver and the china have been set. The best careers are reserved for the students who do exactly what was timely in the youth of their professors, and the tin gods tolerate no *hybris*.

In the spring of 1919, I heard of two teaching openings which seemed to me almost equally attractive. I heard of one through a teachers' agency at the Case School of Science in Cleveland; the other, which was called to my attention by Pro-

THE RETURN TO MATHEMATICS

fessor Osgood of Harvard, was at the Massachusetts Institute of Technology. I do not think that Professor Osgood had a particularly high opinion either of me or of the job, for the previous contributions of the Massachusetts Institute of Technology to mathematical research had been scant, and the department was then almost exclusively a service department to prepare the students for their later mathematical requirements in engineering. However, the postwar rush was absorbing every man who could possibly be considered as a mathematician. Indeed, I had had some hopes that Professor Veblen would take me, as he took Franklin and many others, as a member of the Proving Ground group out of which he was to build the justly famous Princeton department of mathematics; but there were many good candidates, and I was not among his selections.

I called on Professor Tyler, the head of the department of mathematics at the Massachusetts Institute of Technology. He was a small, bearded, bright-eyed man, not a research mathematician, but sagacious and eager for the welfare and the reputation of his department. He appointed me as one of the instructors to handle the new overload, with a possibility of working into a more permanent position if I should make good, but with no promises. He suggested that I should devote myself to applied mathematics. My immediate research in the department happened to be in pure mathematics, but my present happy thirty-three-year association with the Massachusetts Institute of Technology Department of Mathematics and my contact with engineers and engineering problems have given my pure mathematical researches an applied coloring, so that I may be said in some measure to have obeyed the injunction that Professor Tyler laid on me.

At this time, Harvard was involved in a great dispute concerning a Jewish *numerus clausus*. In order to conserve President Lowell's idea of Harvard as a non-sectional institution and the cradle of the ruling class, he proposed to set a certain percentage restriction on the number of Jews admitted. It was understood that it was an administration measure, so that anyone who opposed it risked burning his fingers. My father took an uncompromising attitude against the numerical restriction of the Jewish students, and I am proud to say that when this specific piece of injustice and depreciation came into question, my mother's opinion supported that of my father without any hesitation. All this was happening during the period in which I myself had to look for the firm foundation of my academic career. My sense of belonging to a group which was treated unjustly killed the last bonds of my friendship and affection for Harvard.

I had not been aware at all of anti-Semitic prejudice in my childhood. My parents had many friends who liked and admired them at a certain distance, but they had very few people with whom they felt at liberty to pay unannounced visits, or whom they would have expected to pay unannounced visits to them. I do not believe that this was due to any genuine rejection of my family by the greater part of their Harvard colleagues but much more to a timidity which kept my parents from running the chances of such a rejection.

This was extended to us children. There were very few of the other Harvard children whom I would have been allowed to visit unless such a visit had been organized in advance with the proper amount of interfamily protocol. Thus I was thrown very largely on nonuniversity families for my companionship, and in the end I think this was good.

As to the origin of the family timidity, I think it must have been very mixed. Probably the element contributed by our specifically Jewish race was slight compared with our being new Americans among old Americans, and Westerners among New Englanders. At any rate, it represented a certain degree of aggravation of the relatively isolated position in which we

THE RETURN TO MATHEMATICS

children found ourselves. But all of this is an unimportant element compared with the other features of my situation as a child.

However, by the end of the First World War, I was fully aware of the existence of anti-Semitic prejudice of a most vicious sort. This was a period when it became customary for the friends and faculty advisors of Jewish boys to warn them that their chances of establishing themselves in an academic career were pretty slim. It represents a point of view which lasted for a considerable period but which seems to have gone under in the reassessment of racial attitudes that took place during and after the Second World War.

I have been gratified to see a considerable change of attitude not only toward Jewish scholars in the universities but by Jewish scholars in the universities. With the decline of anti-Semitism, there has been a decline of resentment and fear on the part of the Jewish scholars themselves and a greatly increased possibility for them to participate actively in the problems of the community at large. That this change and maturing has taken place within the circles which I see about me day by day is a matter which is clear to me by the use of my own eyes; and I hope and believe that it is but the counterpart of a phenomenon taking place on a much larger scale.

On the whole, Lowell won the bout in favor of the *numerus clausus*, at least for the period of his own rule. He was defeated, but set up an administrative scheme that enabled him to keep a pretty close watch on the numbers of all Jews not of outstanding talent. I believe that the issue is pretty much dead now after the horrifying example of Nazism and after our more enlightened attitude concerning the right of each individual to employment and to the best education he can manage to get. During the Lowell period, however, those who opposed the president in a matter so close to his heart ran the risk of permanently alienating his good will. After the faculty meetings on the *numerus clausus*, my father was no more able to feel himself in President Lowell's good graces. He felt this keenly sometime later when he hoped for a period of employment by Harvard beyond the statutory age of retirement. He was ultimately denied the privilege and was refused in language that took no cognizance of his long and faithful service to Harvard.

After another New Hampshire summer, but before term began, I found two matters of importance to occupy my attention. I received a call from Barnett, a young mathematician from the University of Cincinnati. As Barnett was working in functional analysis, which was a field in which I aspired to work, I asked him if he could suggest to me a good problem of research. His reply has had a considerable influence on my later scientific career. He suggested the problem of integration in function space. During my first year at the Massachusetts Institute of Technology, I found a formal solution of the problem, which employs some ideas already worked out by the English mathematician, P. J. Daniell, then teaching at the Rice Institute in Texas. However, my first adaptation of the Daniell ideas seemed to me rather lacking in content; so I set out to look for some physical theory that would embody a similar logical structure. I found this in the theory of the Brownian motion. A somewhat similar theory of integration had been discussed by Gâteaux, a young French mathematician who died in the First World War; but his work could not be directly subsumed under that of Daniell and Lebesgue.

Most of my later work in mathematics goes back in one way or another to my study of the Brownian motion. In the first place, this study introduced me to the theory of probability. Moreover, it led me very directly to the periodogram, and to the study of forms of harmonic analysis more general than the classical Fourier series and Fourier integral. All these concepts have combined with the engineering preoccupations of a professor of the Massachusetts Institute of Technology to lead me to make both theoretical and practical advances in the theory of communication, and ultimately to found the discipline of Cybernetics, which is in essence a statistical approach to the theory of communication. Thus, varied as my scientific interests seem to be, there has been a single thread connecting them all from my first mature work to the present.

The other piece of business which I found awaiting me on my arrival in Boston was of a much more worldly nature. The housing and pay of the Boston police force had long been notoriously inadequate, and certain members of the force had stuck their necks out in an effort to have these conditions improved. This led to a threatened police strike. Now there had been a number of such abortive strikes elsewhere at about the same time, and conservative public opinion had begun to be terrified of the possibilities, and to consolidate itself in opposition to the asserted right to strike on the part of those performing vital public functions. Thus there was no difficulty in recruiting a volunteer police force to serve in case the regular police should make their threat good. A friend from the Harvard mathematics department had sent his name in as a candidate for this volunteer police force, and in a moment of misguided patriotism, I followed.

What happened is history, and it made a spurious reputation for Calvin Coolidge, who was Governor of Massachusetts at the time. The regular police struck. Instead of calling out the volunteers to take over the police stations as the regulars left, Coolidge let the city taste twenty-four hours of anarchy and looting before he did anything. This may have been mere indecision or it may have been political sagacity; but whatever it was, it was hard on the shopkeepers whose windows were smashed, and on the nerves and pocketbooks of the public at large. We volunteers received badges and revolvers, and were sent out in pairs to patrol our beats. I was attached to the Joy Street station. The first night I was on duty, there were crowds milling up and down Cambridge Street and Scollay Square and Hanover Street, but we experienced no violence on my beat although a man was killed on a neighboring beat. Later on, I was sent to patrol various streets of the West End. Nothing very exciting happened to me, although I was sent with another recruit to help arrest a wifebeater in a slum near the North Station. I drew my revolver, but it was trembling like the tail of a friendly dog, and I must bless my guardian angel that it did not go off. On another occasion, as I was walking up and down a quiet Jewish slum street, I saw a boy discussing with his comrades some difficulties in an algebra lesson. I interrupted him and put him right, and continued to walk my beat. Sometime afterward, this boy attended the Massachusetts Institute of Technology and became one of my first advanced students in mathematics. I saw him last a couple of years ago at the Carnegie Institute of Technology in Pittsburgh, where he is now a professor.

The net result of the police strike was to make Calvin Coolidge President, to secure the firing of the striking policemen, and to bring in a new police force which was granted most of the demands for which the men of the old police force had sacrificed their careers. As for myself, I was left with nothing but the shame of having acted as the Governor's dupe and strikebreaker.

My arrival at the Massachusetts Institute of Technology meant that I had come safely into port in the sense that I was no longer to be rushed by the problems of finding a job and knowing what to do with myself. When I arrived there, I was one of the large crop of new instructors needed to handle the increased teaching loads which followed in the wake of the First World War. There was no understanding of permanence in my appointment, although I had as good a chance as anyone else to make it permanent if I should prove to be intellectually and emotionally able to make good as a teacher.

The M.I.T. mathematics department itself was then going through a period of transition. Although it was primarily viewed as a service department, there was nonetheless a small nucleus of mathematicians with great scientific enthusiasm who had more or less recently come into the department, and who looked forward to the day when our group might be known as much for its original research and for its training of men capable of carrying out original research as for running interference for the engineering backs.

Among the older men of the department, F. H. Woods had already shown an interest in pure analysis; and E. B. Wilson, who had recently left the department for physics and was destined to leave physics for biostatistical work in the Harvard School of Public Health, represented the great Yale intellectual tradition of Willard Gibbs. Lipka and Hitchcock had been producing for years a certain amount of highly individual mathematical research; this, however, had been off the beaten track, and had a somewhat slender relationship to the type of research favored at other American mathematical schools. The two staunchest supporters of a new policy of research, and the two men who really envisaged what the school might become, were C. L. E. Moore and H. B. Phillips.

Moore was a tall, powerfully built man whose eyesight had only just emerged from a half-blindness because of a displacement of his lenses, and who was to succumb after a few years to a half-blindness resulting from glaucoma. He was kind, intensely loyal to scientific research, and free from any taint of sham. He had studied in Italy before the First World War and had met there an atmosphere of kindness and sincerity which reinforced his own. Italy was then the great home of geometry, and he was accordingly a geometer. Though his field of work was different from mine, he encouraged me with a fatherly interest in my possibilities, which was just what a diffident and awkward young man needed to bring him out. He backed me in the founding of a local M.I.T. mathematical journal, which made it somewhat easier for me to publish my early unorthodox mathematical work.

Professor Phillips, who has officially retired but has not fully withdrawn from teaching, has always seemed to me an utterly timeless figure in the mathematical background of M.I.T. He did not look particularly young when he was young, and he scarcely looks older at the age of seventy. He is a long, loose-jointed Southerner, born in a South where the memory of the Civil War and reconstruction had overshadowed everything else, and he had thus become a skeptic and a bit of a pessimist in a most optimistic and forward-looking way. He was intensely an individual and as fundamentally kind as Moore.

What Moore and Phillips did for me was to discuss their own work with me and to let me discuss mine with them. It must have been very boring for them to receive my ideas in a half-baked form and to suffer under my immature presentation of personal and scientific difficulties. But the great thing was that they listened to me, and that for the first time I had my hopes of becoming a real mathematician reinforced by the confidence of others. Between us we discussed long plans for the future of our department and indeed for the growth of mathematics in the United States. It made me feel more like a man to receive the confidences and the hopes of these men whom I respected, and I became more of a man. Even Professor Moore, who died in 1932, lived to see our department far more than a service department, and indeed one of the constructive research departments of the Institute. Professor Phillips was head of the department for many years after it had assumed its modern functions. What both of these men saw went beyond the wildest dreams they had formed by the end of the First World War.

Within three or four years of my appointment to M.I.T., I began to have accumulated a very considerable body of recognized work. I came to be interested in potential theory, in which I received many suggestions from Professor Kellogg, who was at Harvard at that time. Gradually it began to be clear to me that in the unsolvable cases of the problem of fitting a potential to certain boundary values, there was still a unique potential function fitting these boundary values in a looser sense than that demanded in the literature. Then the question arose, How can one be sure in a particular case that the solution of the generalized Dirichlet problem, as the problem of potential fitting is called, will satisfy the conditions of continuity demanded in classical potential theory?

About that time, a series of papers had appeared by the great mathematician Borel on a different but remotely related subject, called quasi-analytic functions. The innovation of Borel's work at that time was that he brought the problem to depend, not on the size of a number, but on the convergence or divergence of a series. It struck me that my problem of singular points on the boundary of a potential function might well receive its answer in that form instead of in the form of the determination of some particular number, as had been suggested in most of the earlier attempts to solve the problem. At any rate, I sweated the answer through, and my conjecture was correct. With the aid of my Mexican student, Manuel Sandoval Vallarta, who later became a professor at M.I.T. and has since become one of the brightest stars in the firmament of Mexican science, I translated my article into French and sent it to Professor Henri Lebesgue for publication in the *Comptes Rendus* of the French Academy of Science. I followed this course because I had recently seen a series of articles by Lebesgue and a young mathematician named Bouligand, which were getting uncomfortably near to the complete solution of the problem in which I was interested, and which would eliminate the problem from the literature.

It turned out that after I had mailed my article, but before it had been received, Bouligand had submitted a sealed envelope containing a very similar result, to be held by Lebesgue to preserve Bouligand's claim to priority in the field. It was a dead heat between Bouligand and me, and when my article came in, Lebesgue advised Bouligand to let his envelope be opened. The two papers appeared side by side in the *Comptes Rendus*. The results were substantially the same, although I like to think my formulation of the problem was logically a little more complete.

This incident was the start of a friendship between Bouligand and myself which has lasted to the present day; and when, some time afterward, I went to visit him at Poitiers, he made himself known to me on the station platform by exhibiting a copy of my reprint on the subject.

There was a mathematical congress at Strasbourg in the summer of 1920. Although this congress was in some ways unfortunately limited in that Germans were not admitted, I attended it. This was my first opportunity to participate in international mathematics. I worked with Fréchet, who was then professor at Strasbourg, and spent part of my summer vacation at a hotel in the Vosges near where he was staying.

As the result of my work, I participated in two papers of research which were destined to have a certain effect later on. I converted my rather awkward and formal work on integration and function space into a study of the Brownian motion, thus uniting it with the ideas of Einstein and Smoluchowski. This work was an intrinsic stage in developing my later techniques which I have applied in communication theory and in Cybernetics.

The other idea which I developed in my discussions with

THE RETURN TO MATHEMATICS

Fréchet was that of a certain generalized vector space, for which I gave a set of postulates. I soon found out that I had missed the boat by the very narrow margin of a few months, as the theory of the same space had been developed and studied by Banach in Poland. Although we had run almost an even race, I gave up the field later, and left it completely for Banach to open up, as its degree of abstractness struck me as rendering it rather remote from that tighter texture of mathematics which I had found to give me the highest esthetic satisfaction. I do not regret having followed my own judgment in the matter, as there is only a certain amount of work that a mathematician can do in a given time, however he distributes his efforts. It is best for him to do this work in a field that will give him the greatest inner satisfaction.

When I returned to M.I.T. I found that the electrical engineers were beginning to count on me for resolving the very serious logical doubts which were attached to the new and powerful communication techniques of Oliver Heaviside. I was able to make a good deal of progress in this direction, and in the course of this work, I found it necessary to expand the theory of Fourier series and integrals into a more general trigonometric theory covering both. Thus when Harald Bohr of Copenhagen developed his theory of almost periodic functions, I found it to be a field in which I had already developed adequate techniques, and I developed two or three significant alternative approaches to this new subject. The relations between Bohr and myself were and always remained friendly until his death a year and a half ago.

From the beginning of my relationship with M.I.T., I have received loyal backing from it and an understanding of my needs, limitations, and possibilities. I had the opportunity very early to do graduate teaching, and from those early days on, I have collaborated with my younger colleagues and have tried to bring out their intellectual powers. I did not find myself particularly adapted to the niche of the undergraduate teacher. However, the important thing is that I did find that there was a niche in university teaching in which I could function effectively, and this lent me the self-esteem necessary to a successful career.

My undergraduate teaching experience differed so greatly from that at the University of Maine that it was a relief. Perhaps the Maine boys had wanted to play; the Tech boys certainly wanted to work. While there were occasional classroom pranks, they were rare; and the spirit underlying the relations of the professor and the student was one of mutual respect. There were now and then sporadic problems of discipline, but these were so few that they did not form an important part of my relations with my students. Moreover, I could be confident that I would have the backing of the authorities of the Institute in any matter involving my reasonable authority.

At the same time, I learned many lessons. I learned to curb my naturally rapid pace of teaching and to adjust it to students not too much above an average level of performance. I learned that in classroom discipline, the sharp tongue is such an advantage to the professor but such a weapon that it is the part of magnanimity and good sense to use it very sparingly. I learned the trick of handling myself before a student audience, and I sloughed for good all stage fright before a class or, for that matter, before any other audience with serious intellectual purposes.

The year I started teaching at M.I.T. at the age of twenty-five, one of the young ladies who came to our family teas particularly attracted my attention. She was of French background and specialized in French at Radcliffe. She had been brought up before and during the World War in Paris, and was beautiful in a pre-Raphaelite way, with that static beauty which dominates over the beauty of motion in the paintings of Rossetti.

I was greatly attracted to her and spent a great deal of time visiting her and taking her out. She did not like the eternal presence of my much younger brother, and as a result my parents and my sisters took a great dislike to her. I was pursued by them with ridicule, and family ridicule was a weapon against which I had no defenses. I do not know how our interest in one another would have proceeded if it had been left alone. But regardless of that, in the second year of my acquaintanceship with her, she told me she was engaged to another man. I did not take this with good grace, but it was not a graceful situation.

After this, I looked more and more to the Appalachian Club walks for outdoor exercise and social amusements. It had been nearly eight years since I had gone on these walks, and now I found myself more suited in age and in social maturity to those about me. I met several young people and had the chance to discuss many things of interest to us so that I made a definite step ahead in my social development. Nevertheless, I still needed more social contacts and I found these in my parents' teas as before.

About the time that I had met the young woman of whom I have just spoken, I met another young woman who attracted me very much, and if I had not at the time been in the middle of a courtship, I would have had no hesitation in paying court to her. After the breakdown of the other affair and the period that was necessary to rebuild my self-esteem, I began to see her and finally to hope that she might become my wife.

Her name was Margaret Engemann, and she has been my wife now for more than a quarter of a century. My attention was called to her by seeing the same family name in the list of my own students, who were also invited to these teas, and in the list of my father's students in Russian literature. We learned that Margaret and my student, Herbert Engemann, were sister and brother, and that they had been born in Silesia in Germany but had lived in several parts of our Far West. One line of their ancestors went back to Bavaria, and although they looked much alike, Herbert's hair was fair and Margaret's so dark that it was nearly black. They had come to Cambridge from Utah, where they had been students at Utah State College. They were both serious, vigorous young people, to whom I was greatly attracted, and when I later came to know their mother (their father had died many years earlier in Germany), I recognized her as an active and interesting woman of the pioneer type. Margaret shared her mother's definite character, though with a more feminine touch.

On one occasion during the winter of 1921, my family went to their new country place, a farm in Groton, for a little skiing. My parents invited the Engemanns to come along with us. I had taken Margaret out once or twice before this and had enjoyed our companionship very much. My parents considered her an excellent match for me, and were not silent in their approval of my interest in her. However, I felt greatly embarrassed by their obvious reaction in her favor, and my response was to keep away from Margaret for the time being. A courtship that might end in marriage could be only my own and could not represent a decision imposed on me by parental authority. Thus it was not easy for me to show my attentions to Margaret. She has since told me that her reactions to my parents' obvious hints were exactly the same as mine.

On our return from Groton, I began to feel very ill, and it was not long before I came down with an attack of bronchial pneumonia. I was delirious for some days, and during my delirium and convalescence, I expressed the desire to see Margaret again and to talk over our future together. I now felt that she was the wife for me. Yet our courtship and the steps leading up to our marriage did not proceed quickly. I was still confused by my parents' overactive participation in my own affairs. Furthermore, Margaret was soon to leave Boston to take a position as a teacher of French and German at Juniata College in Pennsylvania. Her four years of contact with Juniata have made her a permanent and beloved tradition there.

Margaret shares with me deep emotional roots both in Europe and in America. She was born in Silesia, where she had her early school training; but she came to America with her mother and brothers at the age of fourteen, to share with them memories of that vital part of America, the frontier. Thus she has always combined deep understanding of her mother country and her foster country, and a sincere loyalty to the true interests of both.

From the beginning when Margaret and I have discussed our problems together, she has insisted firmly that I should recognize with honesty what I am, and that I accept my Jewishness with neither boastfulness nor apology. I believe that when I was contemplating marriage, my family had supposed that Margaret would fit rather easily into the somewhat patriarchal pattern of the Wieners, and would serve as a ready instrument for holding me in line. While my parents seemed to have hoped this, I was delighted to find that in fact this was never a possibility. Yet until we were both clear on the matter, we had to wait.

I think that the possibility of marriage had long been in the back of Margaret's mind as well as of mine. We met once at the house of one of Margaret's friends, about halfway between her college and Boston, but on that occasion we were both too involved in our immediate futures to come to an agreement. Yet as time went on it became more and more obvious that we were strongly attracted to one another. I gradually came to see what was never subject to any real doubt: that my parents had taken a great deal too much for granted if they supposed that my marriage with Margaret would mean an indefinite prolongation of my family captivity.

During the years that followed my trip to the Strasbourg mathematical congress in 1920, I visited Europe several times with my sisters and alone, doing a bit of desultory mountain climbing together with some American mathematical friends, and visiting the University of Göttingen in Germany. In 1925, Professor Max Born of Göttingen came to M.I.T. to lecture on physics. It began to seem as if there might be enough interest in my work for me to receive an invitation to lecture at Göttingen. The money was furnished by the newly formed Guggenheim Foundation, which has done so much to help American scholars and artists in all fields. I decided to go to Göttingen in the spring.

With this ahead of me, I felt that I was for the first time in a position to marry at once. Margaret and I had met again at my parents' Cambridge house at Christmas and had decided to marry. There was, however, the difficulty that Margaret's teaching obligations did not end until June, by which time I should be on the other side of the ocean. We tried to see whether there was any possibility of our marrying in Germany at the United States Embassy, but we finally came to the conclusion that this would cause us more effort than it was worth. Finally, we decided to be married in Philadelphia a few days before my departure for Europe, and then to go about our respective jobs until Margaret could come to Germany at the end of term. We spent a few pleasant days of our interrupted honeymoon at Atlantic City, and then parted in New York in a gloom caused to some extent by the fact that we had taken a room in that ancient mausoleum, the old Murray Hill Hotel, and in part by the fact that the play we had chosen to see together was one of Ibsen's gloomiest.

THE RETURN TO MATHEMATICS

However, the period of our separation came to an end, although it seemed it never would, and we met again for the beginning of our European honeymoon at Cherbourg. This was twenty-six years ago, and we were thirty-one. I cannot express how my life these twenty-six years has been strengthened and stabilized by the love and understanding of my partner.

« EPILOGUE »

THIS, THEN, concludes the account of my life from my birth in 1894 until 1926, when I was married at the age of thirty-one. I had joined the staff of the Massachusetts Institute of Technology, and there I have remained ever since.

The present book, apart from its interest to those who have had some continued contact with me and my work, will be read primarily by those who are interested in what is unusual in my career, and the fact that I have been what is known as a child prodigy. There will be many who read it from curiosity, to learn what such a fabulous monster is, and how it views itself. Others will want to find some lessons that they can apply to the education of their own children, or of such other children as may be entrusted to their care. They will ask themselves, and will ask me, some serious questions: Has my career as an infant prodigy been more beneficial or more harmful to me? Would I repeat it if I had the chance? Have I tried to bring up my own children on the same basis, and if I have not, do I wish that I had?

EPILOGUE

These questions are easier to ask than to answer; in fact, one has only one life to live, and the experiment of that life can scarcely have an accurate control. It might be theoretically possible to carry out a controlled experiment with those curious moieties of human beings known as identical twins, but to carry such an experiment out to the bitter end would presuppose a supreme indifference to the development and the happiness of the individual. My father was no such indifferent tyrant. His was anything but a cold nature, and he was firmly convinced that he was doing his best for me. Thus, in the nature of the case, the answer to these questions must be an emotional guess rather than the precise considered verdict of the scientist.

I have tried not to make this volume a *cri du coeur*. It will nevertheless be manifest to the most casual reader that my boyhood life was not all cakes and ale. I worked unconscionably hard, under a pressure which, though loving, was unconscionably severe. With a heredity which of itself would tend to lead to emotional tenseness, I was put through a course of training that was bound to exaggerate this trait, under the impact of another tense personality. I was naturally awkward, both physically and socially; and my training did nothing to alleviate this awkwardness and probably increased it. Moreover, I was intensely conscious of my shortcomings and of the great demands on me. These gave me an unmitigated sense of difference, which did not make it easy for me to believe in my own success.

I was endowed with what was obviously a very real precocity, and with an insatiable curiosity which had driven me at a very early age to unlimited reading. Thus the question of what was to be done with me was one that could not be put off indefinitely. I myself have met more than one able mind which has come to nothing because the ease with which it has learned has insulated it from the discipline of the ordinary school and nothing has been given with which to replace this discipline. It is precisely this rigorous discipline and training which I received from my father, though perhaps in rather excessive portions. I learned my algebra and geometry at so early an age that they have grown into a part of me. My Latin, my Greek, my German, and my English became a library impressed on my memory. Wherever I may be, I can call on them for use. These great benefits I acquired at an age when most boys are learning trivialities. Thus my energies were released for later serious work at a time when others were learning the very grammar of their professions.

Moreover, I had the chance to sit under a very great man, and to see the inner operations of his mind. It is neither family conceit nor filial loyalty which makes me say this. I have lived the life of an active scholar for a third of a century, and I know very well the intellectual mettle of those with whom I have come into contact. My father's work was marred by flights of fancy to which he was unable to give full logical support, and more than one of his ideas has failed to stand the test of later criticism. To be a pioneer in a subject which, like philology, has a very attenuated inner discipline, is to run this risk. My father was a rather isolated worker, an enthusiast, and a man who had come from a different early career. This made his shortcomings almost inevitable; yet his influence in philology is comparable with that of Jespersen, and was an anticipation of the modern school of philologists who see in the cultural history of a language a stronger stream of continuity than in its merely formal phonetic and grammatical development. Both the phoneticists and the semanticists of the present day have come to a position closer to that of my father than to that of most of his contemporaries.

My work with my father may seem to have been an almost unbroken series of clashes, and indeed the clashes were not few. He was a sensitive man, who felt the lack of the general

recognition which he conceived to be his due. He sought for
me to be not only his disciple but his friendly critic and perhaps his continuator. These were impossible roles for even
a mature trained philologist to hold simultaneously; and they
were absolutely out of the question for a half-grown boy.
When I expressed any doubts of his logic, and I had some
sincere doubts, I was berated as an impudent, unfilial child.
Yet I could perceive at the same time the agony of my father
and his need for approval. I knew that he sought for approval
in what he felt was the one quarter in which he could expect it.
Thus my self-protective anger and resentment were not unmixed with pity.

Father was disappointed that his work did not achieve what
he considered and what I consider adequate early recognition.
He was not by any means a failure, nor did he think he was a
failure, either in his intellectual contribution or in the general
frame of academic success. As to the latter, Father reached
and held the rank of full professor at Harvard, and was without any doubt esteemed very highly as a linguist and philologist
of most individual genius. Yet among the very colleagues who
esteemed him, I think there were few who realized that the
position he was taking in the philological world was revolutionary. Neither do I think that, notwithstanding his respect
for his Harvard colleagues, many of them represented to him a
stage of philological learning and sophistication which could
constitute a code whose judgment had any great meaning for
him. Before he had repudiated Germany and Germany had
repudiated him, his heart had been set on a German recognition which was unattainable in the closed German philological
world of that epoch. Even after his break with all things German, I think he still looked toward Europe and hoped that in
some miraculous way, a dove would appear from nowhere with
an olive branch in its beak. I think he could never have looked
forward, except as a dream, to the present state of affairs in

which European scholarship is largely concentrated in America and in which his own point of view, instead of being regarded as a vision of a brilliant eccentric, is accepted and accredited.

Yet the fact that a posthumous success was awaiting him so little as fifteen years after his death can scarcely have mitigated the essential tragedy of his position. And it is possible to be a tragic figure even with an honored position at a great university and a considerable degree of regard among one's colleagues. This position Father had attained, and my mother must be given great credit for taking a brilliant and unworldly man and leading him to that degree of personal success at which he eventually arrived. It was a great success and he knew it. But it was not the position of a re-founder of a science which he deserved and to which he aspired. He had aspired to be Prometheus bringing light, and he suffered in his own eyes the fate of a Prometheus.

From him I learned the standards of scholarship which belong to the real scholar, and the degree of manliness, devotion, and honesty which a scholarly career requires. I learned that scholarship is a calling and a consecration, not a job. I learned a fierce hatred of all bluff and intellectual pretense, as well as a pride in not being baffled by any problem which I could possibly solve. These are worth a price in suffering, yet I would ask this price to be exacted of no man who has not the strength to stand up to it physically and morally. This price cannot be paid by a weakling, and it can kill. That I was a boy not only endowed with a certain intellectual vigor, but also physically strong, made it possible for me to bear the wounds of this Spartan nurture. Before I should even think of subjecting any child, boy or girl, to such a training I should have to be convinced not only of the intelligence of the child, but of its physical, mental, and moral stamina.

EPILOGUE

Even if we take this stamina for granted, it is a special treatment only to be employed where no ordinary treatment is adequate to the needs of the case. With my own children, indications of the need for such a highly specialized procedure have not occurred. At no time have I tried to subject them to a similar training. I cannot say what I should have done if I had found myself faced by the problem that faced my father.

Nevertheless, to confine one's interest only to that part of my development in which my father participated most directly would be to misread the lesson of this book. By the time I had obtained my doctor's degree at Harvard I had completed the usual formal education of the American boy who goes into scholarship. But neither in age nor in sophistication was I ready to take my place in the scholarly world and to earn my living. It is important for me to tell not only how I fell into the rather specialized and stagy-looking life of the infant prodigy, but how I fell out of it and returned to a possible norm. For it is my opinion that this is of interest and importance equal to that of the departure from this norm.

Before I could take my full place as a mature scholar in the world it was necessary for some of the special conditioning which made me to a certain extent an object of show to be replaced by a basis of general experience which must ultimately come to every boy in his teens. I had to learn to study away from the example of my dominating father, and to regulate my affairs among people to whom my record as an infant prodigy meant exactly nothing. I had to become a reasonably competent teacher and to know my assets and limitations in that field. I had to get my hands black in an industrial laboratory and to acquire the satisfaction of working with tools as a member of an active team of men. I had to find out that writing for a living is not done by fits and starts but is a disciplined act which must be repeated for so many hours each day. It was necessary that I should come to see that mathematics was something that dealt with actual numbers and measurements found by observation, and that the results of this mathematics were subject to a critical scrutiny for their accuracy and applicability. And, because I came to maturity in a generation of war, I had to have the knowledge in my own person of what it meant to be a soldier, if not a warrior.

In the career of the average scholar, many of these lessons are learned in the teens and are followed by a period in the twenties when quite as rapid progress is made as I made at an earlier age. This is the more normal procedure and there is much to be said for it. But I hesitate to pronounce dogmatically whether it is better or worse than the alternative procedure which I followed. On the one hand, there have been social difficulties in my nature which not even my belated professional career has eliminated. On the other hand, in this varied period of manifold experiences, my eyes were already open so that I could see and classify and organize in terms of some central principles the mass of individual items that came to my attention. I could come very near the boast that not one of these seemingly desultory years of finding myself was wasted, and that I have integrated them all into a later career centering about a few highly organized principles.

Yet from a contemporary point of view, it must have seemed that I had stepped down from the brilliant glow of publicity belonging to a career of a *Wunderkind* into the half-light of a slightly alleviated failure. I think this interpretation of my career, which would have looked very plausible at about the time I came to the Massachusetts Institute of Technology, is not the true one. I have chosen for the work of my later years the study of communication and communication apparatus. This is a subject with linguistic and philological sides which I have learned from my father, with engineering techniques to which I received my apprenticeship in the General Electric laboratories and at the computing table at Aberdeen Proving

EPILOGUE

Ground, with mathematical techniques stemming from my days at Cambridge and Göttingen, and with the compelling need for a competent vehicle of literary expression which has proceeded from my work on the *Encyclopedia* and the Boston *Herald*. My routine task of assisting a Japanese professor has borne fruit in my teaching in the Orient and my contact with Oriental scholars. Even my exile at the University of Maine, which was a chastisement for me, has proved in the long run to be a salutary chastisement and a true discipline for a man who was to make his living as a teacher and who had the necessity of making his mistakes early when they were of no great seriousness.

This did not result from any particular plan on my part or on the part of my father. The man who wants to work in diverse branches of science must be prepared to take his ideas where he finds them and to use them where they become applicable. Everything is grist to his mill. Indeed, the peculiar advantage of the ex-infant prodigy in science—if he has any advantage and has managed to come through his discipline without major trauma—is that he has had a chance to absorb something of the richness of many fields of scientific effort before he has become definitely committed to any one or two of these. Leibnitz was an infant prodigy, and in fact the work of Leibnitz is precisely the sort of work for which the training of the infant prodigy is peculiarly suitable. The scientist must remember and he must reflect and he must correlate. It does not change the situation in any fundamental way that the field of science has so grown that the scholar of the present day must perforce be nothing more than a half-Leibnitz. The task of scientists is even more essential than it was in Leibnitz's time; and if it cannot be fulfilled with the completeness which seemed possible in the seventeenth century, that part of it which can be carried out is more demanding and less avoidable.

All of this represents a view backward from my later years and not a view forward from my childhood. I began my work early, but my accomplishments did not begin to take their full form until my middle twenties. There was much trial and I went up many false alleyways in going through the maze of life. Yet I doubt that a more single-purposed and unmistaking career would have been better for me in the long run. I do not think that a scientist is at his best until he has learned to draw success from confusion and failure and to improvise new and effective ideas on the basis of procedures which he has begun fortuitously and without purpose. The man who is always right has not learned the great virtue of failure. Intellectual achievement involves a calculated risk and in many cases even an uncalculated risk; but one thing is sure: where nothing is ventured, nothing is gained.

This I should like to say to the administrators of research and education within and without our universities. Theirs is the task of judging the promise and performance of gifted and struggling young men and women, and their decisions may profoundly affect the careers of these young people. The youngsters whom they have to judge must perforce do much of their work in fields in which there do not yet exist any accepted criteria of performance. All true research is a gamble, and the payoff is anything but prompt. A fellowship is a long-term investment in a man, not a sight draft nor a paper collectible twelve months from issue. Creativity cannot be hurried, and Clio takes her time in handing out the awards.

As to the problems of my earlier life which are associated with the fact of my Jewish origin and my discovery of it, these have evaporated with time. I have found in my wife's attitude support for a definite course of conduct and security in that course of conduct. As I have said, this is to sublimate the problem of prejudice directed against a group to which I belong into the problem of prejudice against undervalued groups in general. In addition, whatever temporary recrudescences anti-Semitism may show, it has ceased to be a really important factor in the environment in which I live, and to a large extent in the country as a whole. Among those places at which anti-Semitism is at a low level and has ceased to be an important factor in our daily lives, the Massachusetts Institute of Technology stands at the top.

This decided lessening of anti-Semitism is the result of many factors. The shame of Hitler's anti-Semitism has cut deep into the spirit of most Americans and it is no longer fashionable or even tolerable to adopt such a discredited spirit. Furthermore, the Jews, like many other immigrant groups, have developed a new generation which has grown up with American speech and American mores and which no longer combines the prejudice-forming factors of different dress, language, and background with the prejudice against religious differences and the Jewish group in particular. The struggle for emancipation from the ghetto does not have any very great emotional content for those to whom this emancipation is an old story. Yet the battle against prejudice is never won, and it must be fought on every front where it appears.

All in all, it has come out pretty even in the end. The question of one's social awkwardness looks pretty small after one has gone through the vicissitudes of fifty-eight years of life and has found oneself reasonably able to cope with them. The early start I have had does not appear to me to have impeded me from showing a period of productivity continuing reasonably late, and has greatly increased the level at which I started this productivity. Thus it has added years to my useful life.

I certainly do not look back upon my career as one blighted by my earlier experiences, nor do I feel any particular self-pity for having been "deprived of my childhood," as the

EPILOGUE

jargon goes. That I have arrived at this degree of equanimity is due in particular to the love, advice, and criticism of my wife. Alone, I should have found it difficult and probably impossible of accomplishment. Yet it has been achieved. And now, with increasing age, I find that the image of myself as an infant prodigy has been obliterated in the minds of my acquaintances as well as in my own mind. The question of the success or failure of my adolescence and postadolescence has become unimportant to me as to everybody else through the larger issues developed during my career as a working scholar.

« INDEX »

A

Aberdeen, Maryland, 254-263
Aberdeen Proving Ground, 254-263, 294-295
Adams, Eliot Quincy, 105, 131
Adams, Henry, 53
adolescence, 102-179
Adventures of a Young Naturalist in Mexico, 83-84
Agassiz Museum, 64
Agassiz School, Cambridge, 38
Albany, New York, 248-254
Albee, Ernest, 149-150
Alcott, Louisa, 108
Alexander, J. W., 257
Alexandria, New Hampshire, 37-38
Alger, Horatio, 108
Alger, Philip, 255
Alice in Wonderland, 37
American Mathematical Society, 224
ancestry, 8-30
anglophobia, 184-185

Anthology of Russian Literature (Wiener), 85
anti-Semitism, 131, 153-156, 271-274, 297
Appalachian Mountain Club, 168, 283
Arabian Nights, 37, 63, 65
Arithmetic (Wentworth), 46
Arnold, Benedict, 201
Audacious, 219
Autobiography (John Stuart Mill), 76
Ayer High School, 67, 92-101

B

Babbage, Charles, 8
Bahaists, 96
Baker, Louise, 139
Baker, H. F., 190
Baldwin, Miss (principal of Agassiz School), 38
Balkan Wars, 181
Banach, Stefan, 224, 281
Bangor, Maine, 242
Bangor General Hospital, 242

INDEX

Barnett, Isaac A., 274
Bartlett, F. C., 187-188, 202
Basic English, 13
Bauer, Frau, 205-206
Bavarian Court Library, 198
Beard, Daniel Carter, 64
Beaumont, Dr. (army doctor), 59
Beith, Ian Hay, 218
Belmont, Massachusetts, 141
Bennett, A. A., 257
Bentham, Jeremy, 70
Berle, A. A., 131, 136, 137-138, 177
Berlin, Germany, 14
Bernstein, Felix, 211
Birkhoff, G. D., 230-231, 255
Black Beauty, 95
Blackwood's Magazine, 218
Bliss, J. I., 256, 257
Bôcher, Maxime, 58, 181, 232
Boer War, 56
Bohr, Harald, 209, 281
Bohr, Niels, 191, 194
Borel, Félix Édouard, 279
Boring, Miss (Student at University of Maine), 241-242
Boros, Max, 194
Börne, Ludwig, 123
Boston, Massachusetts, 29
Boston *Herald*, 265-267
Boston police strike, 275-276
Boston Public Library, 29, 30, 36, 59, 107
Boston Sunday *Herald*, 135
Boston *Transcript*, 15, 219
Boston University, 29, 59

Bouligand, C. D., 279-280
Bowdoin Prizes, 202
Brattleboro, Vermont, 54
Bray, Hubert, 257
Bridgewater, New Hampshire, 157
Bridgman, Percy, 166
Broad, C. D., 270
Bromley, Kent, 57
Brown, Frank, 95
Browning, Robert, 21
Bruce, H. Addington, 73, 119, 158 fn.
Busch, Wilhelm, 56, 65
Büschen, Fräulein, 204
Butler, Nicholas Murray, 222
Butler, Samuel, 71-72, 167
Byelostok, Russia, 10, 11

C

calf love, 99-100
Cambridge, England, 183-203
Cambridge, Massachusetts, 31, 32-33, 38, 58, 60-61
Cambridge Magazine, 188
Cambridge Philosophical Society, 201
Cambridge University, 185-203, 218-221
Camp Devens, 262
Cannon, Walter B., 59, 113
Carlton College, 171
Carnegie Foundation, 114
Carnegie Institute of Technology, 276
Carroll, Lewis, 37, 195, 270

« 300 »

INDEX

Castle, W. E. (geneticist), 59
Catalpa Farm, 40-41
Cayuga Lake, 151
Central Park, New York City, 49, 51
Central Park Zoo, 49
Century Magazine, 157
Chao Yuen Ren, 229
Cherbourg, France, 287
Chesterton, G. K., 48, 263
Child, Francis, 29-30
childhood, 31-101
Cincinnati, 217
Civil War, 26
Clark University, 177
Cologne, Germany, 52, 54, 197
Columbia, Missouri, 28
Columbia University, 220-225
Communists, 235
Comptes Rendus, 279, 280
Conrad, Joseph, 194
Cook, Dr. F. A., 114
Coolidge, Calvin, 275, 276
Coolidge, Julian Lowell, 232, 233
Cooper, James Fenimore, 107
Copley Theater, Boston, 241
Cornell University, 140, 143-152, 215
Cosmopolitan Magazine, 84
courtship, 282-287
Crothers, Samuel McCord, 83
Cudjo's Cave (Trowbridge), 108
Cushman, H. E., 109
Cybernetics, 8, 275, 280

D

Daniell, P. J., 274
de Broglie, Louis, 194
Demos, Raphael, 227, 237
Deutsches Museum, Munich, 200
Dewey, Admiral George, 39
Dewey, John, 222
Dickinson, Lowes, 194
Disney, Walt, 81
Docent Lectures, Harvard University, 212, 223, 230, 231, 269
Dolbear, Katherine, 119 fn., 177
Donlan (grocer at Ayer), 97
Dostoevsky, Feodor Mikhailovich, 74
Drew, Daniel, 26
Drummer Boy (Trowbridge), 108
Duffy, Mary, 36
Duffy, Rose, 36, 82
Dumas, Alexandre, 85, 108, 169
Du Maurier, George L. P. B., 50
Dupriez, L., 221

E

education, 43-47, 66-68, 75-77, 92-108
Edwards, General C. B., 267
Eger, Aqiba, 10
Einstein, Albert, 191, 194, 200, 280
Eliot, Charles William, 125

INDEX

Eliot, T. S., 220
Ellinger family, 23-24
Elliott, Richard M., 204
Emerson, Ralph Waldo, 256
Encyclopedia Americana, 248-254
Engemann, Herbert, 284
Engemann, Margaret, 283-287
Englischer Garten, Munich, 200-201
Esperanto, 13
Estes and Sons (publishers), 85
European travel, 51-57, 181-217, 218-221, 280, 287
Evans, G. C., 131, 132
eyesight deficiency, 74-76

F

family background, 8-30
Father and Son (Gosse), 72
Fay, Professor "Tard," 104
First Hundred Thousand (Beith), 218
"First Men in the Moon," (Wells), 84
First World War, 160, 163, 169, 207, 214, 216-217, 218-221, 236-237, 243, 247-263
Fisher, Bud, 114
Folin, Otto, 58-59
Fort Slocum, 259
Fourierist community, 19
Foxboro, Massachusetts, 40-41, 84
Frankfurt, Germany, 54
Frankfurter, Felix, 138

Franklin, Phillip, 257-258, 271
Fréchet, Maurice, 280, 281
Freud, Sigmund, 31
Funktionentheorie (Osgood), 265

G

Galileo, 73
Galsworthy, John, 194
Gardner, Mrs. Jack, 57
Gâteaux, R., 274
General Electric Company, 247-248, 258, 294
Gibbs, Willard, 8, 277
Gill, B. P., 258
Gödel, K., 193
Goebbels, Joseph, 205
Gosse, Edmund, 72
Göttingen University, 201, 204-216
Grant, Ulysses S., 26
Graustein, W. C., 258
Green, Gabriel Marcus, 258-259, 262, 264
Green Letters, 229
Greenacre, Maine, 96
Greene, Graham, 183
Grimm's Fairy Tales, 97
Gronwall, T. H., 257
Groton, Massachusetts, 284
Guide of the Perplexed (Maimonides), 144

H

Haldane, J. B. S., 12
Hall, G. Stanley, 177

« 302 »

Hambourg, Mark, 56
Hamburg, Germany, 216
Hamburg America Line, 217
Hammer, 205
Hammond, W. A., 149
Hardy, G. H., 183, 189, 190, 194, 209, 216
Harper, Miss (secretary), 85
Harper's Magazine, 157
Harte, Bret, 17
Harvard, town of, Massachusetts, 86-87
Harvard Mathematical Society, 211, 231, 233
Harvard Philosophical Library, 176
Harvard Regiment, 236, 238, 243
Harvard Union, 130, 196
Harvard University, 29, 125-140, 164-179, 228-237
Harvard University Graduate School, 113, 125-140, 152, 215
Harvard University Medical School, 59
Haskell, Dr. (oculist), 75
Haskins, Professor C. N. (Dartmouth College), 262
Hattori, U., 156, 228
Havana, Cuba, 15
Heaviside, Oliver, 281
Hebraische Melodien (Heine), 169
Heine, Heinrich, 70, 169
Heisenberg, Werner, 191, 194
Henderson, Lawrence J., 166

Henry, O., 243
Henry of Prussia, Prince, 74
Heredity in Royalty (Woods), 166
Hilbert, David, 209, 210, 215
History of Yiddish Literature (Wiener), 85, 145
Hitchcock, F. L., 277
Hitler, Adolph, 205, 215
Holland-America Line, 51
Holt, E. B., 165
Holy Rollers, 41
Homer, 170
Horace, 170
Hospital of the Holy Ghost, 32
Houghton, Cedric Wing, 131, 137, 138-139
Howard, Hermann, 33, 43
Hugo, Victor, 108
Humboldt Library, 63, 65
Huntington, Edward Vermilye, 167-168, 181, 232
Husserl, Edmund, 210
Hutchinson, J. I., 150, 190
Hyslop, W. W., 142

I

Illustrated London News, 219
Imperial College of Science and Technology, London, 205
infant prodigy, 117-119, 131-139, 177-178, 215, 288
Institute for Advanced Study, Princeton University, 181
Israel, National State, 9

INDEX

J

Jackson, Dugald, 142
Jaffrey, New Hampshire, 32
James, Henry, 110
James, William, 109-110, 141, 164, 165, 176
Jerusalem Talmud, 10
Jespersen, Otto, 122, 290
Jewish family structure, 9, 11-12, 50-51
Jewish origin, problem of, 144-149, 152-156, 285, 296-297
Jewish Publication Society, 146
Joffre, Joseph Jacques, Marshal, 217
John the Orangeman, 74
Johns Hopkins University, 241
Jones, Miss E. E. C., 195
Jordan, David Starr, 199-200
Josephine (French maid), 33
Jourdain, P. E. B., 269-270
Journal of Philosophy, Psychology and Scientific Method, 192-193
Joyce, James, 249
Judaism, 8-9, 12, 153-154
Judson, Adoniran, 238
Jungle Book (Kipling), 34
Juniata College, 285

K

Kahn, Henry (grandfather), 22, 23, 86

Kahn, Mrs. Henry (grandmother), 86
Kaltenleutgeben, Germany, 55
Kansas City, Missouri, 19-21
Karner, Edward, 224
Keller, Helen, 74
Kellogg, Oliver, 279
King, Everett, 107, 243
King, Harold, 107, 141
King of the Golden River (Ruskin), 83
Kingsley, J. S., 111, 112, 130
Kipling, Rudyard, 30, 54, 155, 169
Kittredge, Dora, 33
Klein, Felix, 209-210, 231
Kraus, Karl, 54-55
Kropotkin, Prince Pëtr, 57
Krotoschin, Russia, 10, 11

L

Lafayette, Marquis de, 89
Lake Pontchartrain, 18
Lambert, F. D., 110
Landau, Edmund, 209
Lawrence, Massachusetts, 266-267
Lawrence Scientific School, Harvard University, 232
Leavitt, Laura, 93, 95, 101
Lee family, 36-37
Leibnitz, Gottfried Wilhelm von, 8, 109, 193, 295
Lebesque, Henri, 279, 280
Letters from John Chinaman (Dickinson), 194

INDEX

Levy, Hyman, 205-206
Leyland Line, 181
Lincoln, Abraham, 26
Lipka, Joseph, 277
Littlewood, J. E., 190, 209
Liverpool, England, 15, 57
London, England, 56-57, 182-183, 220
London *Times*, 249
Loomis, F. W., 255
Lowell, Abbott Lawrence, 125-126, 175, 236, 244, 271, 273

M

Maggy, "The Buttonbreaker," 62
Maimonides, Moses, 10, 143-144, 155
Mainz, Germany, 54
Maloney, Hildreth, 62
marriage, 286-287
Mason, A. E. W., 108
Massachusetts Institute of Technology, 132, 173, 189, 244, 258, 271, 276-286, 288, 294, 297
Material for the Study of Variation (Bateson), 111
Mathematical Society, Gottingen University, 210-211
Mathematician's Apology (Hardy), 189
Max and Moritz (Busch), 55-56, 65
Maxwell, James Clerk, 8
McTaggart, J. M. E., 195, 196

Meaning of Meaning, 188
Mendelssohn, Moses, 10, 14, 133, 156
Mercer, J., 190
Merriam-Webster *Dictionary*, 30
Messenger of Mathematics, 190
Metivier, Renée, 43-44
Mikado (Gilbert and Sullivan), 35-36
military service, 258-263
Mill, James, 69-71
Mill, John Stuart, 68-71, 72-73, 76
Milyukoff, Pavel Nikolaevich, 97-98
Minsk Gymnasium, 12
Modern Algebra (Bôcher), 181
Modern Painters (Ruskin), 83, 170
Modern Symposium (Dickinson), 194
Monist, 270
Moore, C. L. E., 277-278
Moore, G. E., 165, 195-196, 216
Moral Science Club, 188, 195
Moreh Nebukim (Maimonides), 144
Morize, Major, 244
Mount Passaconaway, 246
Mount Washington, 237
Munich, Germany, 183, 197-201, 201-202, 203, 204
Monroe, Walter, 79

INDEX

Münsterberg, Hugo, 164, 167, 231
Murray, Patrolman, 78
Murray Hill Hotel, New York City, 224, 286
Mursell, Jim, 237
Muscio, Bernard, 188
Muss-Arnoldt, W. (Assyriologist), 59-60
Mysterious Island (Verne), 84

N

Natural History (Kingsley), 63, 111
Natural History (Wood), 34
Nesbit, Evelyn, 108
New England Conservatory of Music, 29, 30
New Machiavelli (Wells), 195
New Orleans, Louisiana, 15, 18
New York City, 10, 48-51, 220-225, 286
New York State Guard, 252-253
New York State Library, 250
New York State Museum, 250
New Yorker Magazine, 134-135, 138, 178

O

Officers Training Camp, Plattsburg, 237-239
Ogden, C. K., 188
Old Mill Farm, 86-91, 92, 96, 100, 102, 107, 113, 167
On a Balcony (Browning), 21

Orono, Maine, 228, 237, 239-243
Osgood, W. F., 231-232, 245, 271
Out on a Limb (Baker), 139

P

Palmer, G. H., 164-165
Parker, C. P., 110
Peabody School, Cambridge, 66, 78, 93
Peano, G., 179
Pearl, Raymond, 241
Peary, Robert E., 114
Peck, Annie, 20
Pedagogical Seminary, 178
Perry, Ralph Barton, 165, 225, 236
Phi Beta Kappa, 116-117
Phillips, H. B., 277, 278-279
Phoutrides, Aristides Evangelos, 227
Piscataqua River, Maine, 96
Plattsburg, New York, 237
Poems (Rosenfeld), 85
Poincaré, Jules Henri, 230-231
Polytechnicum, Berlin, 14
Pope, Alexander, 70
Poritzky, Hillel, 258
Principia Mathematica (Whitehead and Russell), 191, 193, 201, 223
Proceedings of the Cambridge Philosophical Society, 191, 212
Projective Geometry (Veblen and Young), 181

R

Radcliffe College, 30, 33, 228, 241, 282
Rand, Dr. Benjamin (librarian, Harvard University), 176
Ransom, W. R., 104, 112
Rattray, F. C., 166-167
reading, early, 63-66
Reid, Mayne, 63, 107-108
Republic (Plato), 149, 172
Reserve Officers Training Corps, 243-245, 252
Revere Beach, Massachusetts, 114
Rhines, George E. (Editor, *Encyclopedia Americana*), 248, 251, 254
Rice Institute, 132, 257, 274
Richards, I. A., 188
Ring and the Book (Browning), 21
Ritt, J. F., 257
Robertson, Helen, 74
Rockwood, Ray, 78
Rogers, Homer and Tyler, 97, 100
Röntgen ray, 59
Rosenfeld, Moritz, 55, 85, 146
Rotterdam, Holland, 52, 197
Royce, Josiah, 164, 165-166, 171, 176
Ruckmich, Christian, 151
Ruskin, John, 83, 170
Russell, Bertrand, 165, 171, 176, 179, 181, 183, 190, 191-195, 196, 200, 201, 208, 216, 217-218, 219-220, 222, 229, 230, 270

S

Sage Fellows, 150
St. Catherine's College, Cambridge University, 187
St. Joseph, Missouri, 22, 23
St. Martin, Alexis, 59
St. Nicholas Magazine, 38-39, 63, 64, 100
Sandwich, New Hampshire, 140-141, 170
San Francisco, California, 20
Santa Claus myth, 81-82
Santayana, George, 164, 194
Schaub, E. L., 151
Schiller, F. C. S., 177
Schmidt, Karl, 171
Schofield, W. H., 30
Schrödinger, Professor Erwin, 191, 194
science fiction, 84-85
Scottish Ballads, 29
Scribner's Magazine, 157
Second World War, 101, 132, 257
Sessions, Roger, 131, 137, 139
Shaker village, 89-90
Shaw, George Bernard, 167
Sheffer, Dr. H. M., 240-241
Sidis, Boris, 131, 133-134
Sidis, W. J., 131-136, 137, 138, 168, 177, 178
Silver Lake, New Hampshire, 245

INDEX

Smith, W. Benjamin, 28-29
Smoluchowski, M., 280
Somerville, Massachusetts, 31
Southard, E. A., 166
Spanish-American War, 35, 39
Spencer, Herbert, 103
Spinoza, Benedict, 109
Stevenson, R. L., 63
Strand Magazine, 108
"Stray Leaves from My Life," 15
Struwwelpeter, 65
Studies in Synthetic Logic, 212
Swift, Jonathan, 169
Szasz, Otto, 211

T

Tamarack Cottage, 140
Taylor, Dr. (physician), 82
Taylor, Isaac, 97
Tenniel, Joseph, 195
Thackeray, William Makepeace, 169, 234
Thaxter, Ronald, 129
Théorie des equations integrales (Fréchet), 265
Théorie des equations integrales (Volterra), 265
Theory of Ignorance, 96
Thilly, Frank, 140, 143, 152
This Week Magazine, 135
Thompson, Benjamin, 201
Three Scouts (Trowbridge), 108

Through the Looking Glass, 37
Titanic, 169
Tolstoy, Count Leo, 70, 72, 74, 85
Top of the World, 237
Treasure Island, 63, 95
Trowbridge, J., 108
Tsanoff, R. A., 151
Tsing Hua University, 242
Tufts College, 102-113, 177
Tulane University, 28
Turing, A. M., 193
Twain, Mark, 17, 243
Tyler, H. W., 271

U

Unitarian Sunday School, 83-84
University of Maine, 237, 239-243, 282, 295
University of Mexico, 106
University of Missouri, 28, 29, 143
University of Puerto Rico, 254
University of Warsaw, 13
Utah State College, 284

V

Vallarta, Manuel Sandoval, 279
Vanderbilt, Cornelius, 26
Veblen, Oswald, 181, 254, 255, 257, 259, 271
Verne, Jules, 84, 107
Vienna, Austria, 54-55

INDEX

W

Wade, C. S., 103, 151
Warsaw, Poland, 12
Way of All Flesh (Butler), 71
Wells, H. G., 84, 85, 195
Whiskey Ring, 26
Whiteface Mountain, 246
Whitehead, Alfred North, 201, 220
White Mountains, 227
Widder, David, 258
Widener Library, Harvard University, 175-176
Wiener, Adele (aunt), 50
Wiener, Augusta (aunt), 49, 50
Wiener, Bertha (sister), 82, 83, 86, 98, 158, 183, 199
Wiener, Bertha (Kahn), (mother of author), 22, 26-28, 34, 50-51, 53, 58-59, 68, 82, 120, 144, 146, 198, 203, 225, 292
Wiener, Charlotte (aunt), 42, 49-50
Wiener, Constance (sister), 36, 39, 51, 76, 83, 90-91, 158, 162, 183, 199, 241, 245, 253, 259, 262, 264
Wiener, Fritz (brother), 98, 158-161, 162, 199
Wiener, Jake (uncle), 49, 51
Wiener, Leo (father of author), 12-22, 28-30, 34, 36, 37, 40, 43, 46-47, 51, 52-53, 61, 62, 67-74, 75-77, 85, 86, 94, 120, 122-124, 145-146, 198-199, 217, 234-236, 289, 290-292
Wiener, Moritz (uncle), 50
Wiener, Margaret (wife of author), *see* Engemann, Margaret
Wiener, Olga (cousin), 41, 42-43, 49, 51, 84, 144, 225
Wiener, Solomon (grandfather), 10-11
Wiener, Freda (grandmother), 41-42, 50, 225
Willard, Winn, 79
Wilson, E. B., 277
Winthrop, Massachusetts, 113-114
Wirth, Jacob, 241
Wolfson, Harry, 259
Woods, Frederic Adams, 166, 172
Woods, F. H., 277
Wren, Dean F. G., 112
Wyman, Arthur D., 63-64

Y

Yenching University, 242
Yosemite Valley, California, 20
Young, J. W., 181
Youth's Companion, 79

Z

Zamenhof, Lazerus Ludwig, 13
Zangwill, Israel, 56-57, 154, 183, 196
Zionism, 9

THE MIT PAPERBACK SERIES

Computers and the World of the Future, edited by Martin Greenberger (originally published in hard cover under the title **Management and the Computer of the Future**)   MIT 1

Experiencing Architecture by Steen Eiler Rasmussen   MIT 2

The Universe by Otto Struve   MIT 3

Word and Object by Willard Van Orman Quine   MIT 4

Language, Thought, and Reality by Benjamin Lee Whorf   MIT 5

The Learner's Russian-English Dictionary by B. A. Lapidus and S. V. Shevtsova   MIT 6

The Learner's English-Russian Dictionary by S. Folomkina and H. Weiser   MIT 7

Megalopolis by Jean Gottmann   MIT 8

Time Series by Norbert Wiener (originally published in hard cover under the title **Extrapolation, Interpolation, and Smoothing of Stationary Time Series**)   MIT 9

Lectures on Ordinary Differential Equations by Witold Hurewicz   MIT 10

The Image of the City by Kevin Lynch   MIT 11

The Sino-Soviet Rift by William E. Griffith   MIT 12

Beyond the Melting Pot by Nathan Glazer and Daniel Patrick Moynihan   MIT 13

A History of Western Technology by Friedrich Klemm   MIT 14

The Dawn of Astronomy by Norman Lockyer   MIT 15

Information Theory by Gordon Raisbeck   MIT 16

The Tao of Science by R. G. H. Siu   MIT 17

A History of Civil Engineering by Hans Straub   MIT 18

Ex-Prodigy by Norbert Wiener   MIT 19

I am a Mathematician by Norbert Wiener   MIT 20